Devils Hole Pupfish

America's National Parks Series

Series Editor, Michael Welsh, University of Northern Colorado

America's National Parks promotes the close investigation of the complex and often contentious history of the nation's many national parks, sites, and monuments. Their creation and management raises a number of critical questions from such fields as archaeology, geology and history, biology, political science, and sociology, as well as geography, literature, and aesthetics. Books in this series aim to spark public conversation about these landscapes' enduring value by probing such diverse topics as ecological restoration, environmental justice, tourism and recreation, tribal relations, the production and consumption of nature, and the implications of wildland fire and wilderness protection. Even as these engaging texts cross interdisciplinary boundaries, they will also dig deeply into the local meanings embedded in individual parks, monuments, or landmarks and locate these special places within the larger context of American environmental culture.

Death Valley National Park: A History
by Hal K. Rothman and Char Miller

Grand Canyon: A History of a Natural Wonder and National Park
by Don Lago

Lake Mead National Recreation Area: A History of America's First National Playground
by Jonathan Foster

Coronado National Memorial: A History of Montezuma Canyon and the Southern Huachucas
by Joseph P. Sanchez

Glacier National Park: A Culmination of Giants
by George Bristol

Big Bend National Park: Mexico, the United States, and a Borderland Ecosystem
by Michael Welsh

Devils Hole Pupfish: The Unexpected Survival of an Endangered Species in the Modern American West
by Kevin C. Brown

Devils Hole Pupfish

The Unexpected Survival of an Endangered Species in the Modern American West

Kevin C. Brown

UNIVERSITY OF NEVADA PRESS | *Reno & Las Vegas*

University of Nevada Press | Reno, Nevada 89557 USA
www.unpress.nevada.edu
Copyright © 2021 by University of Nevada Press
All rights reserved
Cover illustration © Joseph Tomelleri
Cover design by Iris Saltus

All photos, images, and illustrations by author unless otherwise noted.

LIBRARY OF CONGRESS CATALOGING-IN-PUBLICATION DATA
Names: Brown, Kevin C. (Kevin Conor), 1984– author.
Title: Devils Hole pupfish : the unexpected survival of an endangered species in the modern American West / Kevin C. Brown.
Description: Reno ; Las Vegas : University of Nevada Press, [2021] | Includes bibliographical references. | Summary: "The Devils Hole pupfish is one of the rarest vertebrate animals on the planet; its only natural habitat is a ten-by-sixty-foot pool near Death Valley, on the Nevada-California border. Isolation in Devils Hole made the fish different from its close genetic relatives, but as the book explores, what has made the species a survivor is its many surprising connections to the people who have studied, ignored, protested, or protected it" —Provided by publisher.
Identifiers: LCCN 2021010749 | ISBN 9781647790103 (paperback) | ISBN 9781647790110 (ebook)
Subjects: LCSH: Devils Hole pupfish—Conservation—History. | Rare fishes—Conservation—Nevada—Devils Hole. | Devils Hole (Nev.)
Classification: LCC QL638.C96 B76 2021 | DDC 597.16809793/34—dc23
LC record available at https://lccn.loc.gov/2021010749

The paper used in this book meets the requirements of American National Standard for Information Sciences—Permanence of Paper for Printed Library Materials, ANSI/NISO Z39.48-1992 (R2002).

FIRST PRINTING

Manufactured in the United States of America

Contents

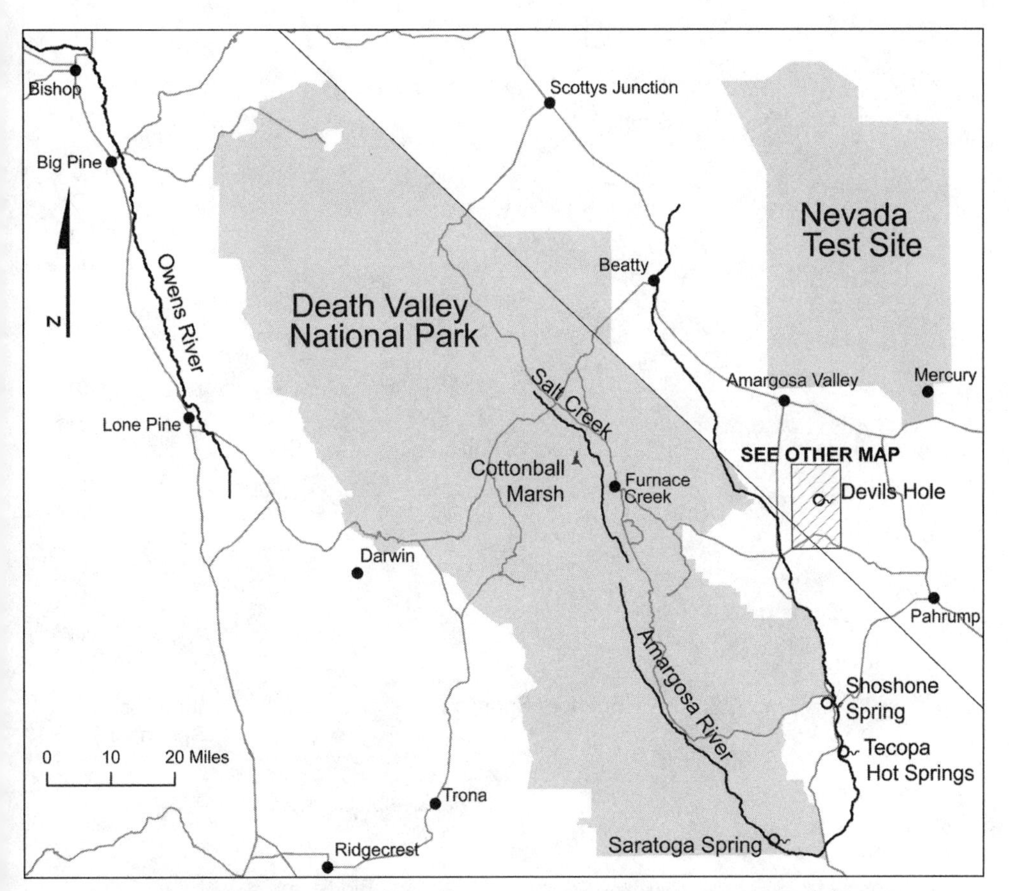

FIGURE 0.1. Death Valley National Park and environs, showing principal roads and towns. Rivers and springs that once contained or still contain pupfish are also shown. Map by author with public-domain data.

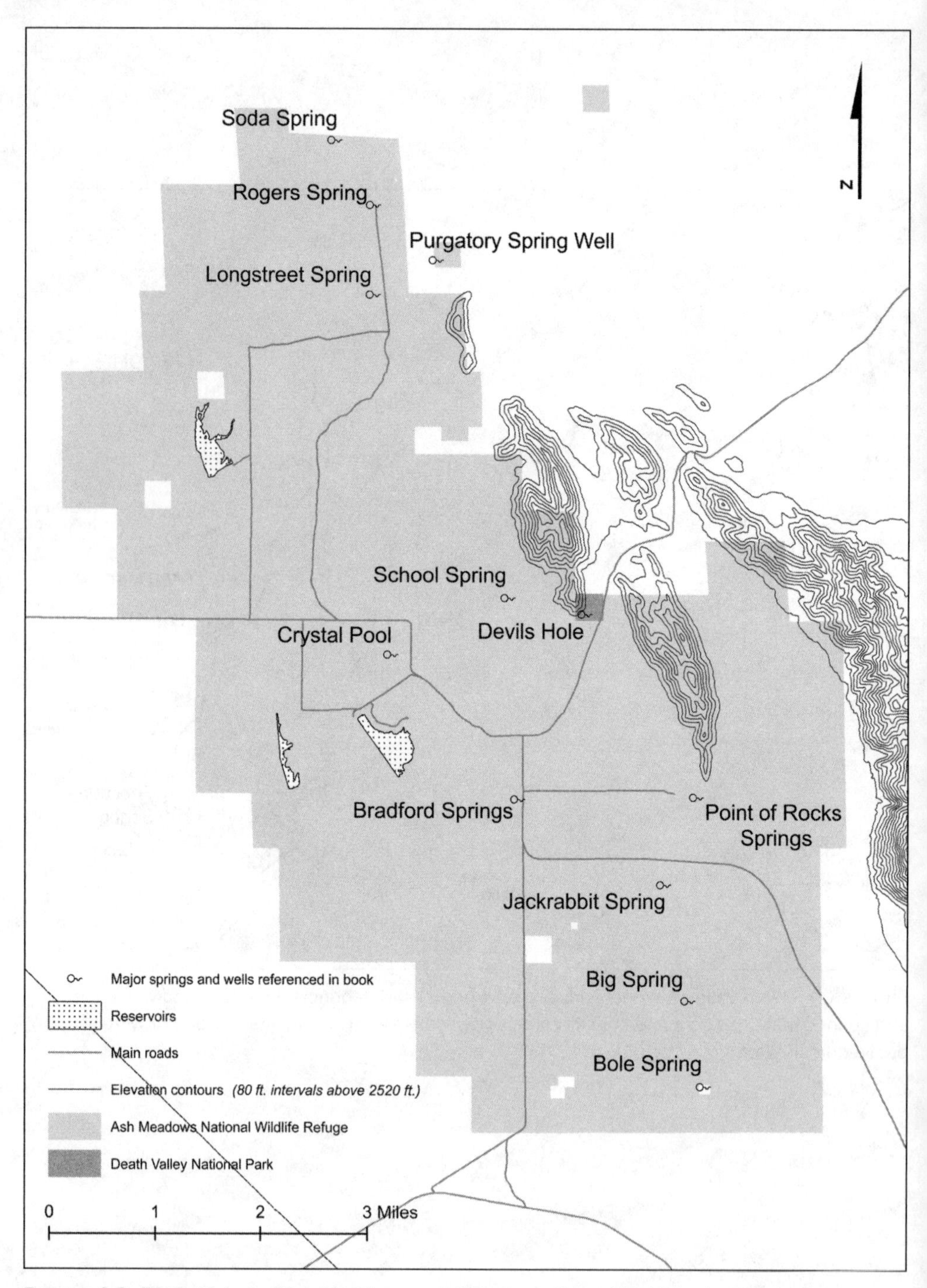

FIGURE 0.2. Devils Hole and Ash Meadows area. Map by author with public-domain data.

Introduction

Surviving the Meteor

Sixty thousand years ago, the roof of a water-filled cavern collapsed just east of Death Valley in what is now southern Nye County, Nevada. Nineteenth-century Anglo American visitors, channeling a cosmology that saw dark forces in both geological oddities and the wilderness, named the place "Devils Hole."[1]

This roughly funnel-shaped feature is, improbably, a window into a vast subterranean lake.[2] When you stand on the platform at the top of the Devils Hole fissure and look down to the pool, 45 feet below, you are staring into an aquifer, a rare vantage point. Water remains a constant 92 degrees Fahrenheit in Devils Hole. One researcher scuba diving there said it felt like returning to the womb.[3] U.S. Geological Survey divers descended to a depth of 436 feet but saw neither bottom nor any evidence of the two teenage divers who disappeared into the abyss in 1965.[4] Devils Hole, for all practical purposes, is bottomless (fig. 0.3).

Deserts are defined by their aridity. But Devils Hole is on the eastern edge of a relatively lush patch of the Amargosa Desert called Ash Meadows. Along a series of faults that run on a northwest–southeast line near Devils Hole, a network of more than a dozen large springs and numerous seeps discharges some 10,000 gallons of water every minute, producing oases in the desert that stretch across some 50,000 acres.[5] Another way to think about Ash Meadows is as an island, though instead of being surrounded by water, it is bounded by land that sorely lacks it. Trapped on the island of Ash Meadows, at least twenty-nine kinds of animals and plants have evolved into species, subspecies, or varieties found nowhere else in the world. Ash Meadows is the Galápagos of the Intermountain West.[6]

This book is about one of these species, an inhabitant of the strange pool on the edge of the Ash Meadows oasis, a one-inch-long, twitchy blue fish: *Cyprinodon diabolis*, the Devils Hole pupfish (fig. 0.4). Despite the unknown

FIGURE 0.3. Devils Hole, as it can be seen today from Death Valley National Park's wire-encased viewing platform.

depth of Devils Hole, the pool is just 10 feet wide by 60 feet long at its surface. The pupfish live only in a small zone near the top of the water column, where invertebrates and algae (i.e., pupfish food) are found. They spawn exclusively in an even more restricted area: a chunk of rock that scientists call the "shallow shelf," wedged into the southern third of the pool and barely below the waterline. The entire species, which at times has had a population of fewer than 40 observed fish, could barely stock a suburban dental office aquarium. The pupfish's only natural home is Devils Hole, and biologists consider the fish to have the smallest habitat for a vertebrate species in the entire world. Even within the species-rich island of Ash Meadows, the Devils Hole pupfish is rare and unique, a fact that helped make it a founding member of the U.S. endangered species list when it was created in 1967.

Arriving sometime after Devils Hole became open to air and light—though exactly when is subject to lively debate—the Devils Hole pupfish was shaped by its isolation in an extreme environment. The water in Devils Hole, in addition to its high temperature, has very little dissolved oxygen. Direct sunlight does not even reach the pool in winter. As a result, more than half of the total annual energy available in Devils Hole comes from plants, insects, and dust; a minority share of energy comes from algae actually growing there.[7] Because it evolved in this energy-poor, high-temperature environment, the Devils Hole pupfish is smaller and less aggressive than other pupfishes (yes, there are others), and unlike these relatives, it almost never even has pelvic fins. In other words, by pupfish standards, *C. diabolis* is both tiny and wimpy.[8]

Yet the Devils Hole pupfish is a tenacious survivor, enduring not just a harsh physical environment, but also recent American history. Nineteenth-century Anglo-Americans often regarded deserts as places to avoid or navigate quickly, an attitude in sharp contrast to the Newe (Western Shoshone) and Nuwuvi (Southern Paiute) peoples that have made the land around Ash Meadows a homeland. The group that gave Death Valley its dour name—a party of easterners seeking their fortunes in the California gold fields—even passed by Devils Hole on December 23, 1849, and at least two members enjoyed a swim in the pool.[9] Despite the aversion to deserts, by the 1870s, the prospect of extracting precious metals—and wealth—made the Amargosa Desert home for a mining industry, eventually earning Devils Hole the nickname "miner's bathtub" from workers living nearby.[10]

FIGURE 0.4. Preserved specimens of Devils Hole pupfish (*Cyprinodon diabolis*) in the collection of researcher Carl Hubbs. Photograph courtesy of Carl Leavitt Hubbs papers, Special Collections and Archives, University of California, San Diego.

The twentieth century brought new uses to the desert around Devils Hole. The U.S. military reserved vast swaths of Nevada and California for airfields, bombing ranges, and a destructive new test site far from the prying eyes of citizens and spies, while the National Park Service (NPS) promoted tourism and expanded its holdings at its desert parks, including Death Valley National Monument in 1933. (The monument was expanded and redesignated as Death Valley National Park in 1994.) These two very different categories of federal lands sometimes intersected at Devils Hole. In 1951, just a year before President Truman added Devils Hole to the Death Valley monument, the government began testing nuclear weapons north of Devils Hole at the Nevada Test Site. Some blasts caused water in Devils Hole to slosh back and forth. Ultimately, more than 1,000 nuclear weapons were detonated.[11]

Recent human history has done more, however, than just swirl around the Devils Hole pupfish while it remained safely ensconced in its remote habitat. Instead, the pupfish has become repeatedly entangled in some of the most important changes in how Americans understand, exploit, and protect the West. For this reason, this book is in large measure about how people have wrestled with the existence and meaning of the pupfish since the 1890s. Scientists, federal and state officials, ranchers, politicians, developers, and even the U.S. Supreme Court have all weighed in. These engagements have produced a changing scientific classification of the Devils Hole pupfish; debates over the aims of the national park system; an extensive battle over water rights and the power of the federal government; and a continuing struggle to manage a species with a very small population.

And yet, despite its turbulent recent history and extreme habitat, the pupfish remain—an act of survival that is a true wonder. It took me years to realize that this modern persistence was, in addition to being a minor miracle, a conundrum in need of explanation. Across a century during which humans have radically altered the deserts, springs, and atmosphere of the American West, and as many endemic species have gone extinct—including some of the Devils Hole pupfish's close genetic relatives and geographic neighbors—how exactly has the Devils Hole pupfish persisted?

This book is my answer to that question, an effort to get to the bottom of why there are still pupfish living in Devils Hole. Each chapter provides a partial answer: The pupfish got a scientific name—not every species does; many remain unstudied. The pupfish happened to live in what was, until the 1960s, an economic backwater—a now-extinct fish species in Las Vegas was less fortunate. The pupfish found a way inside the national park system, administered by a public agency with good intentions (if a short attention span) and considerable authority to protect resources it considers important; other species do not. And on and on. Together, the book's six chapters argue that this species persisted across the twentieth century and into the present not by being isolated but by becoming enmeshed in the human institutions of the modern U.S. West. This does not mean that things have always gone smoothly—people sometimes disagreed about how to act, and the pupfish has not always "behaved" in ways that people would like or expect. Still, the pupfish survived by unwittingly becoming cosmopolitan, incapable of being separated from us and our institutions.

Persistence is not the framework we often use for thinking about rare species. Instead, we are typically preoccupied with the causes of their endangerment and extinction.[12] There is good reason for this preference. Put simply, a lot of species are in trouble. Based on extinction rates currently being observed in nature, scientists worry we are at the outset of the sixth mass extinction in Earth's history. The five previous mass extinction events in the last 500 million years resulted from a variety of changes, including ecology and atmospheric composition as well as climate swings. But they have in common a stunning outcome: at least 75 percent of all species on the planet perished in a relatively short time (in geologic terms). There is some disagreement on how to extrapolate into the future the extinctions observed over the last few hundred years, and a central uncertainty in calculating current extinction rates is how many species have recently gone extinct before even being named by scientists. Still, there is a growing fear in the scientific community that the scope of extinctions we are witnessing may ultimately reach the order of Earth's most recent mass extinction event: the meteor that caused the demise of the dinosaurs.[13] This time around, however, *we* are the meteor.

Human effects on biodiversity are so numerous that we might use the plural instead: humanity as *meteors*. Overhunting, the introduction of invasive species and diseases, land clearing for agricultural or industrial development, impounding water for dams, and pollution have all contributed to extinctions. Often, multiple forces are at work, piggybacking off one another. Climate change, meanwhile, is creating novel ecological conditions, further compounding these and other threats to many species. In spring 2019, a United Nations–backed international panel of scientists released a damning report that detailed the ways that humans have degraded the environment. The study suggested that as a result upward of one million species are at risk of becoming extinct, many within decades. Slowing or halting this tidal wave, the group warned, will require a rapid "fundamental, system-wide reorganization across technological, economic and social factors, including paradigms, goals and values."[14] Mass extinction appears to be at our doorstep.

This book is written in full awareness of the range of threats to biodiversity—and the chilling and impoverished future we face if we do not make the dramatic changes to our society and economy as described by the U.N. group. But it is precisely because of this extinction crisis, our concern for it, and the apparent difficulties in slowing or stopping it, that the pupfish's story of persistence is important. We need to investigate the conditions under which some species have endured, especially when their survival seems surprising and, at times, quite tenuous.

Conservation biology—a field explicitly concerned with the preservation of biodiversity—has its own method for evaluating the likelihood of survival for any (often endangered) species over a given time period, an approach called "population viability analysis," or PVA. By combining information about population trends, life history, and habitat into a mathematical model, biologists can assess the risk to a species based on proposed management actions, likely future environmental conditions, or low population sizes.[15] A PVA model created for the Devils Hole pupfish, published in 2014, suggested that the species faced greater than a 25-percent probability of extinction within twenty years.[16] Survival for the pupfish appears tenuous indeed.

My approach to studying pupfish survival is quite different from that of conservation biology. While PVA models can be useful, they do not explain (or propose to explain) how a species has managed to survive long enough to even have this elaborate technique applied to it. In short, how did the pupfish even make it to 2014? The first record I have found of a person (a budding scientist, no less) worrying that the Devils Hole pupfish might become extinct comes from a letter written in 1937, more than eighty years ago and just seven years after the fish was described as a unique species.[17]

Essentially, people have been worried about the extinction of the Devils Hole pupfish ever since they knew to call it by the scientific name *Cyprinodon diabolis*. And yet, it is still here. Somehow, the species has threaded a very narrow needle through American history that other rare and endangered species have not. I think it is worth trying to understand why—and that methods very different from a PVA are required.

In this book, I use the tools of history to bring into the picture the ways that humans—including scientists and their models—have shaped the survival of the pupfish. The story is rooted in the detective work of locating, examining, and carefully reading through boxes and boxes of material in archives, museums, libraries, government offices, and people's homes, as well as sorting through computer files and conducting interviews. For this project, I also devoted a fair amount of time simply exploring Death Valley and Ash Meadows by foot and car to better understand the landscape I spent my days reading about.[18] Like journalists, perhaps historians' closest disciplinary cousins, our stories are hemmed in by the specifics of what our documentary or human sources can confirm. Yet after uncovering all the memos, reports, data logs, letters, and other sources we can, the major task is figuring out how they fit together and what they mean.

Because the pupfish has survived through its relationship with humans—us—this book explores the history of some of the most important institutions, rules, laws, and cultures that shape the modern West: the U.S. Bureau of Land Management, state water permitting, the Endangered Species Act, and the practices of the ecological sciences, to name just a few. For me, one of the joys of learning about the pupfish's history, and I hope one of the benefits of reading this book, is that it sheds light on places and processes that extend far beyond Devils Hole. If you have never heard of the pupfish—or are not sure you need to know more about its particular circumstances—do not panic. One of the points of looking at such an extreme kind of "case study" is not just to understand its own peculiarities but to see what an unusual situation can reveal about issues we think we know well.[19] To better understand the normal, we must look at the weird.

This story raises challenging questions for us as we live through the era of the sixth extinction and look to the future. Since human actions now have affected the entire globe, to find that a species' persistence in the past depended not on staying isolated from people but by becoming closer to our systems of knowledge production, land tenure, and resource use is comforting and perhaps instructive. It may be that other species need to be closer to us in order to "save" them. Yet, as we will see through the backlash in

Nevada to a consequential water rights case, the costs of making room for the pupfish were not evenly distributed.

Another question for the future has to do with whether we can expect the level of commitment made to the pupfish to be extended to all the other species currently and likely to become vulnerable in the era of the sixth extinction. This question is an old one. At one point in 1970, when the pupfish was threatened by groundwater pumping, the Park Service and other federal agencies put together an ambitious plan for its rescue. The plan raised eyebrows in Washington: "Such a crash program may well save the pupfish," wrote one senior wildlife official, "but what of the other 80+ species also on the list?"[20] Today, with more than 1,600 species on the U.S. threatened and endangered species list, the question remains.

A Note on Usage

"Pupfish" is the common name given to members of the genus *Cyprinodon* living primarily in the United States West, Mexico, and the Caribbean. One species in this group, the Devils Hole pupfish, *Cyprinodon diabolis*, is the principal subject of this work. Throughout, I use "pupfish" to refer to the Devils Hole pupfish specifically, unless the context makes clear that I am describing pupfish as a genus.

There are sometimes questions about the "correct" spelling of this place-name—especially whether it should be written "Devil's Hole" or "Devils Hole." Conventions in language change over time, and this is true for Devils Hole. When scientists began visiting this place in the late nineteenth century, they often referred to it as "the Devil's Hole." Later, others shortened the name to just "Devil's Hole." And by the 1970s and 1980s, those writing about the pupfish increasingly did not include the apostrophe either, adopting a convention common in scientific writing as well as a policy of the U.S. Board on Geographic Names. Throughout, I write "Devils Hole" and "Devils Hole pupfish." To avoid distraction for readers, I have also removed the apostrophe from "Devils" in quoted material.

Devils Hole Pupfish

CHAPTER 1

What's in a Name?

Humans are naming machines. Our ancestors peered into the night sky and turned a galaxy of distant stars into crabs, dippers, and archers. Marketers used a name to heroically transform a boxy, two-door car from Eastern Europe into the lovable Yugo. And before 1950, biologists came up with no fewer than three different monikers for a group of tiny desert fishes: pumpkin seed, pursy minnow, and pupfish. Only the last of these stuck; it came to a researcher after he noticed how the fish often chased each other—just like puppies.[1]

We are also classifiers, adept at ordering the relationships between those things we name.[2] Oat Bran tastes different from Cap'n Crunch, but both belong in the cereal aisle. Libraries are monuments to classification, whether organized by the Dewey Decimal or the Library of Congress systems. The natural world represented perhaps our species' first opportunity to name and classify. One suspects, for example, that our ancestors found it important to be able to communicate whether a "predator" or potential "prey" lurked nearby.

Natural scientists have turned the classification of the Earth's living things into a full-time job, comprising a central part of a field called systematic biology. A physical manifestation of scientists' commitment to classifying can be found—somewhat surprisingly—near the Suitland, Maryland, stop on the green line of Washington, D.C.'s metro. Just over a quarter mile from the platform is a Smithsonian Institution facility called the Museum Support Center, a massive, squat building set back behind a high fence from a busy six-lane highway. Inside one of its five football-field-sized storage "pods" are some 6 million fishes from all over the globe, making it the largest assemblage of its kind in the world.[3]

This Smithsonian facility is not an aquarium; it is a necropolis: all of its fishes are dead. Most specimens, as scientists call the fish, are preserved in alcohol, labeled with reference numbers, and placed in glass jars and tanks. Because of the tremendous quantity of the highly flammable liquid

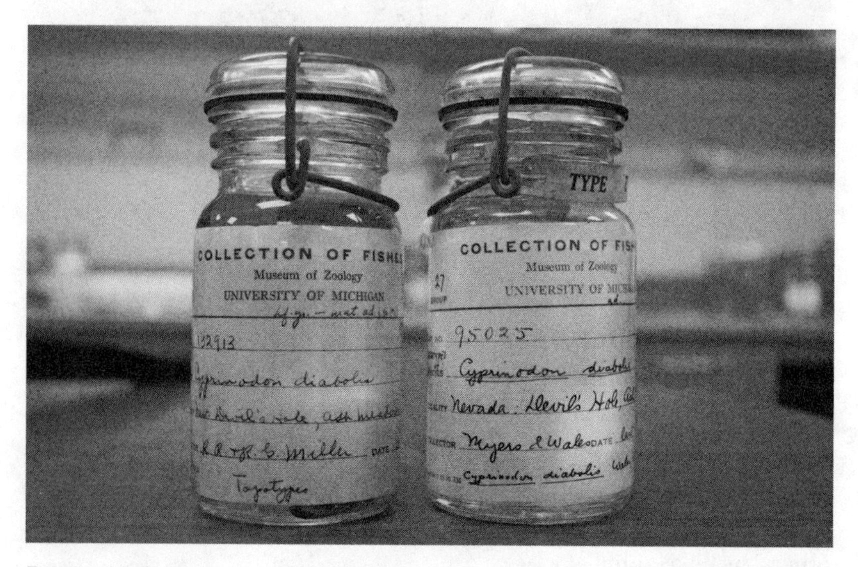

FIGURE 1.1. Specimen jars of Devils Hole pupfish held at the University of Michigan Museum of Zoology. On the left are specimens collected by Robert R. Miller; on the right are specimens collected by Joseph Wales.

required to preserve them, the facility is equipped with a state-of-the-art fire suppression system. One measure taken to contain a potential blaze is the compartmentalization of the pod into a maze of sour-smelling rooms with the collection organized on giant, library-style shelving. On one of these shelves sit several jars containing fish from Devils Hole, including the first ten individuals first taken from the habitat by scientists in 1891 (fig. 1.1).

What makes the jars at this Smithsonian facility so amazing is more than the bizarre sight of rooms full of pickled fishes, though indeed, when I visited, I found that part pretty wild. Rather, the collection's significance is that it exists not to demonstrate what scientists already understood about nature when they bottled the fish but just how little they knew. The Smithsonian's natural history collections—it also has birds, mollusks, and just about all other kinds of nonmicroscopic life—are different from a stamp collection, where the goal is to accumulate *known* "specimens," whether it is an 1847 five-cent Benjamin Franklin, or a misprinted 1918 twenty-four-cent airmail stamp. Scientists collected the fishes at the Smithsonian, by contrast, because it was *not* clear where in the collection of life they should go.

Especially during the nineteenth and early twentieth centuries, biologists made the collection, examination, and storage of animals, including fishes, the critical technique for defining (or redefining) species, charting their distributions, establishing their relatedness to other species, and unraveling the mysteries surrounding the evolution of life on this planet. The pupfishes, just one group of fishes among many others, got wrapped up in this massive scientific undertaking to understand the diversity of life by looking at preserved bodies.

How scientists classify species is not academic. While the methods used to make such judgment were (and are) highly specialized, and the audience for the publication of their determinations very small, the classification of populations can have enormous consequences—especially for whether we ignore, exploit, or protect those creatures. As historian Peter Alagona has observed, "Conservation history... is not merely a political struggle over things that exist in nature; it is a perennial competition to prove the existence and define the very nature of those things that are the focus of such struggles."[4] How we group and differentiate animals determines whether and how we choose to conserve them—or attempt to conserve them—and is why this book starts with the naming and classifying of the Devils Hole pupfish.

Beginning in the 1890s, the way that the fish living in Devils Hole got caught up in the grand movement to classify life has played a crucial role in its long-term survival. More directly, the pupfish survived the twentieth century, in part, because of how it was named and classified. Over a period of decades, and after three ichthyologists examined the pupfish around Ash Meadows, the Devils Hole pupfish was confirmed as a unique species, a finding based on a particular set of scientific practices in the field, in the laboratory, and in publishing. The specific uniqueness of the pupfish served as the rationale for its protection in the national park system in 1952, as well as its place on the first U.S. endangered species list in 1967. And its classification continues to be an important justification for its protection. To answer the question of how the pupfish persisted across the twentieth century, we must start with the first people who intentionally killed them and dragged them out of the desert in jars.

Federal Fish

On March 20, 1891, a young Californian named Theodore Sherman Palmer visited Devils Hole. He described it as a "noted" spring, forming a "deep rift in the bluish limestone." Palmer then scooped up ten "small fish" from the pool.[5] This was the first scientific collection of what we have come to know as the Devils Hole pupfish. I found Palmer's diary entry about the episode in the archives at the Huntington Library near Pasadena, California. It was a somewhat anticlimactic moment—Palmer devoted just a paragraph to his visit to Devils Hole. While I had driven more than two hours through rush-hour traffic to see this diary entry, visiting Devils Hole was not, it seems, especially noteworthy to him.[6]

Still, Palmer fulfilled his central scientific responsibility: these specimens have been well preserved, thanks to the antiseptic properties of alcohol and to generations of curators and archivists. When I visited the Smithsonian Museum Support Center on a muggy September afternoon several years ago, Palmer's fish were doing fine. I popped open jar #46107 and with tongs pulled out the alcohol-soaked, paper identification tag used by Palmer, confirming that the enclosed 125-year-old fish were from Devils Hole.

Palmer's Devils Hole trip was a milestone in the history of the pupfish, not only because it was the first collection made from the site but also because of who employed him. Just twenty-three-years old at the time, he was a "field agent" in a section of the U.S. Department of Agriculture called the Division of Economic Ornithology and Mammalogy. Palmer worked for the federal government.

From the perspective of the twenty-first century, this does not immediately surprise: the federal government today is intimately involved with science and conservation. The National Science Foundation funds basic ecological research, and federal agencies protect animals and plants in parks and refuges as well as through the administration of the Endangered Species Act of 1973. Grizzly bears, island foxes, alligators, and pupfish have all been beneficiaries of such federal attention. Much of this book—along with many others—is about the consequences of the federal government's relationship with animals. Exploring the context around Palmer's collection from Devils Hole reminds us that the federal relationship with wildlife stretches back many decades before the establishment of these familiar institutions. During the late 1800s and early 1900s, the government also played an important role in collecting, describing, and cataloging the species that decades later it would begin protecting through park policy and endangered-species law. It is not just that the federal government included the Devils Hole pupfish in Death Valley National Monument or on the endangered species list but that research by federal agencies actually helped bring the population to science as a classified, named species. Since the 1890s, the pupfish has been a federal fish.

The Death Valley region is hardly the breadbasket of America. And fishes are not typically considered an agricultural product. So how Palmer, an employee of the U.S. Department of Agriculture, ended up collecting fish there is not self-evident. Even more curiously, the explanation for how Palmer found himself at Devils Hole starts with birds.

In the mid-1880s, a newly formed group called the American Ornithologists' Union sought to accurately map the ranges and migratory patterns of the nation's avian fauna. Overwhelmed with the enormity of this undertaking—the project is still ongoing in the twenty-first century through citizen science projects such as the Cornell Lab of Ornithology's eBird app—the group lobbied for federal support, proposing to legislators that accumulating this data would reveal the role birds played in controlling insects that damaged farm crops.

The Ornithologists' Union succeeded in garnering a small appropriation for a new division within the Agriculture Department's Bureau of Entomology. The resulting Division of Economic Ornithology—Palmer's eventual employer—represented a small start to the federal wildlife program. Over subsequent decades and through successive reorganizations and name changes, it morphed into the Bureau of Biological Survey and eventually into today's U.S. Fish and Wildlife Service, which oversees more than 560 national wildlife refuges and administers the Endangered Species Act. But in the late 1880s, this was all in the future. At that time, the division was small, claiming only a $10,000 annual budget and a mustachioed, twenty-nine-year-old physician from upstate New York named C. Hart Merriam for its first chief.[7]

Merriam had an early start as a natural scientist and in government. As a sixteen-year-old, in 1872, he collected specimens for a federal survey of the Yellowstone region.[8] After taking charge of the division at the Department of Agriculture, Merriam, especially early in his tenure, showed a profound lack of interest in the economic part of his bureau's title. "Economic ornithology" be damned—what fascinated Merriam were the forces that governed the distribution of plants and animals, what scientists today call the study of biogeography. Merriam led his first investigation of flora and fauna in northern Arizona in 1889. From a study of the San Francisco Peaks north of Flagstaff, he developed the concept of "life zones" to characterize the changes in plants and animals he found as he moved up and down the mountains. Descending from the highest elevations, above the tree line, Merriam passed through zones of tundra, spruce, fir, and pine forests, before reaching the plants of the Colorado Plateau's desert floor.[9] Merriam did not invent the idea of life zones—credit for a similar concept goes to Alexander von Humboldt decades earlier—but he argued that shifts between zones depended on temperature, and he demonstrated the concept with a compelling, U.S.-based case study.[10] Death Valley, with its "close proximity of precipitous mountains and deep desert valleys," Merriam wrote, was the perfect place to make a second study of these principles and uncover previously undocumented species.[11]

Merriam's group had more than an innovative concept. They also had an established method for studying these "life zones": the survey party. Starting in the 1840s, the federal government began conducting surveys of the West to reconnoiter railroad routes, delineate territorial boundaries, and map geological features and watercourses. These surveys moved quickly across wide areas, and biological collections were often secondary to economic and political objectives: extending the American empire in the West.[12] But Merriam's group, by contrast, made collecting specimens and recording natural history information the central mission of its surveys. In the field, workers moved slowly, collecting many specimens of the same species at different localities in a region, carefully documenting changes in vegetation, elevation, and temperature. And to appreciate how and why the distribution of plants and animals changed, it was also necessary to identify those organisms they encountered. Surveys like Merriam's depended on classifying and naming.

The Death Valley Expedition organized by Merriam ran from December 1890 through the summer of 1891, employing thirteen scientific personnel from agencies that included the Agriculture Department's Botany and Entomology Divisions, the U.S. Geological Survey, and the U.S. Signal Service (a forerunner of the National Weather Service). These workers, along with packers and cooks, often split into smaller groups to cover more territory as they crisscrossed Death Valley, Ash Meadows, the Eastern Sierra Nevada Mountains, and even as far as Las Vegas and Utah's Virgin River (fig. 1.2).[13]

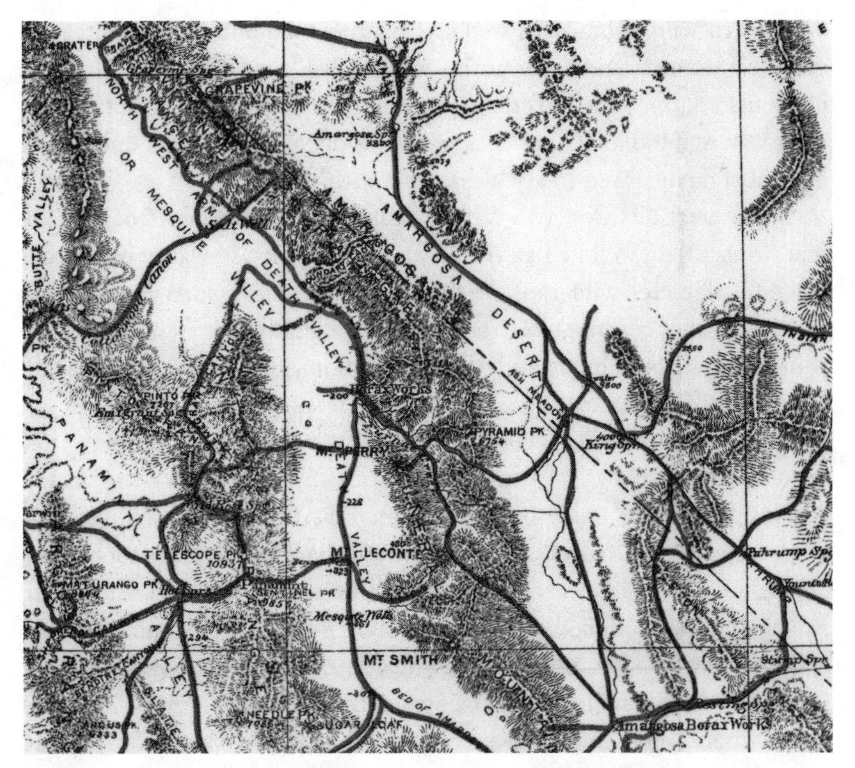

FIGURE 1.2. Detail of map of Death Valley region, with overlaid lines showing the routes traveled by Merriam, Palmer, and other members of the 1891 Death Valley Expedition. From *The Death Valley Expedition: A Biological Survey of Parts of California, Nevada, Arizona, and Utah, Part II*, North American Fauna No. 7 (Washington, D.C.: GPO, 1893), reproduction courtesy of the University of Nevada, Las Vegas Libraries Special Collections.

The type of work done by members of the expedition is difficult to classify (ironically), because, as historian Robert Kohler has noted, scientific fieldworkers required "a range of skills from the arcane and bookish to the handy and practical" in a world that frequently divided these abilities into different jobs.[14] As today, workers in the late nineteenth century were segregated into jobs where they used either their hands or their heads to earn a paycheck. Expedition members cut across this divide: they were expected to be crack shots (both for collecting specimens, and for supplementing their diets); good taxidermists for preparing specimens; willing to endure weeks in the backcountry; knowledgeable in the care of stock animals; able to organize complex schedules; knowledgeable about the likely habitats of different wild animals; as well as able to correspond with scientists in distant museums and laboratories and publish their findings. The work made field naturalists into crosses between cowboy, hunter, secretary, and professor, as if Indiana Jones became deeply dedicated to collecting lizards instead of racing Nazis to mythical antiquities.

The expedition traveled through a landscape quite different from the one that we visit today. While Death Valley National Park is now over 90 percent federally protected wilderness, in the 1890s the whole region was a working landscape. That is, though Merriam's group sought to document the natural world—its plants, animals, and their distribution—they relied on, interacted with, and observed a lot of people living and working in the desert.

Two groups were especially important to the expedition. First, they encountered Native Americans whose ancestors had been living in the desert for hundreds, if not thousands of years. Several periods of Indigenous occupation in the Death Valley region stretch back more than 10,000 years, during which the area varied substantially in its climate, flora, and fauna. Archaeological evidence suggests that ancestors of the Numic-speaking Nuwuvi (Southern Paiute) and Newe (Western Shoshone), groups that today live in the region, migrated to the area within the last 800 years.[15] By the 1870s, when American surveyors passed through Ash Meadows, the area had become a borderland between these two groups, with the surveyors reporting a "small band of about fifty men, women, and children, composed of renegade Shoshones and Pah-Utes [Paiutes], together with a mixture of these two tribes."[16] Merriam's expedition also apparently met members of both groups during their visit to Ash Meadows in March 1891.[17]

More than just witnessing the arrival of the Death Valley Expedition, however, the Indigenous residents of Ash Meadows also materially aided the group and its scientific mission. T.S. Palmer, the young expedition member who collected the first fish from Devils Hole, recorded in his diary that Paiute residents stopped by the expedition's camp in Ash Meadows and donated three as-yet-unnamed species of fishes to Palmer and his colleagues, having collected them from springs, as well as recently dug irrigation ditches, in the area.[18] While Palmer may have captured the first pupfish from Devils Hole himself, at other springs in Ash Meadows Native Americans played a crucial role in literally bringing these species to science.

Palmer acknowledged their role in his diary and also consulted them about directions to other springs in the area, yet no Indigenous names appear on the Smithsonian jars as collectors of record.[19] They were written out of science. In fact, the pupfish collectors do not even appear to have been asked what they called the fish, reflecting in microcosm the broader history of violence toward, and marginalization of, Indigenous people on western landscapes. Among the expedition photographs taken at Ash Meadows, one is particularly telling. It shows Palmer receiving a haircut from a Paiute woman and is captioned, "Prof. T. S. Palmer being barbered by a barbarian barber."[20]

While the Native Americans in the Death Valley region saw the landscape as a place to live, Anglo-American newcomers considered it a place from which to extract valuable minerals—gold, silver, borax—destined for expanding capitalist markets. The Death Valley Expedition also relied on these new residents. The first Anglo-Americans reached the floor of Death Valley in the winter of 1849–1850, but not until after the Civil War did miners and settlers see the region itself as a source of wealth rather than an obstacle on the way to distant riches. In the late nineteenth century, a range of boomtown mining ventures rose and fell around the region, including at Panamint City, Lookout, Chloride Cliff, Lee, and near Furnace Creek.[21] It was from the boomtowns and ranches built to serve the Death Valley mining economy that expedition members restocked supplies and where they rendezvoused.[22] After Palmer met an Ash Meadows resident, George Watkins, while camped in Furnace Creek, he wrote in his diary that he had "learned a good deal about the country from him" and as a result decided that the expedition should itself visit and camp in Ash Meadows.[23]

Despite the presence of Native Americans and Anglo miners, the region was still mysterious enough in the urban West that these government biologists attracted enviable amounts of press attention during the expedition. Both the San Francisco *Examiner* and *Chronicle* sent reporters with Merriam's crew during the spring of 1891. Palmer's visit to Devils Hole even made it into the *Examiner*, with a reporter describing the "small carp like" fish living in the habitat and mentioning efforts to find the bottom of the hole by dropping rope into the pool. The rope hit a rock at 70 feet, which Palmer's party took to be the bottom. "Most of the springs" in the desert were "most peculiar," the paper noted.[24]

While peculiar springs and their inhabitants generated copy for newspapers, they were not the centerpiece of the Death Valley Expedition's work. Merriam and company collected about 1,000 reptiles and amphibians, 1,000 birds, 6,000 mammals, 4,500 insects, and 25,000 plants.[25] With fishes, the group was less prolific: today, the Smithsonian has 105 pickled fish in its collection from the Death Valley Expedition, with about half of these specimens taken from Ash Meadows.[26] Palmer and his colleagues did make collections from springs as far north as the Owens and Pahranagat Valleys and from the Amargosa River in the south, but the group also missed an opportunity to collect fishes at several notable places, including at Salt Creek—today the easiest place for visitors to observe pupfish in Death Valley National Park.

Despite these omissions, the expedition succeeded in putting desert fishes and many other kinds of animals and plants on the map, charting in literal and figurative routes for later scientists to retrace and then explore further. And, while preserving fishes in alcohol or trapping small mammals may not seem like auspicious beginnings for the persistence of animals living in isolated habitats, by making such collections the expedition transformed them not only into objects that could be studied but also—perhaps—conserved. "What we save, when we save species today," geographer Elizabeth Hennessy notes, "are not only animals, but specific ways of understanding them forged through histories of scientific exploration and collection."[27] Yet classifying the pupfish—as well as all other life, including the Galápagos tortoises that Hennessy writes about—depended on more than the collection of specimens. It also required knowing what to do with them once they arrived in the laboratory.

• Found Identical

One of the things I learned by hanging out with ecologists at the University of California, Santa Barbara, where I wrote parts of this book, is that they spend a lot more time indoors than one would guess. Sure, they spend days, sometimes weeks or even months, outdoors collecting data, but most of the year they are inside, in front of a computer or at a lab bench making sense of the information they have brought back from "the field." The same was true of nineteenth- and early-twentieth-century taxonomists—the name for those concerned with classifying and naming living things. Investigating particular habitats was necessarily required to collect animals and plants, but the analysis of those specimens occurred in laboratories far from their point of acquisition.[28]

For the classification of the fishes from the Death Valley Expedition, the division between collecting and classifying, between the desert and the lab, was even sharper, since no one on the expedition even worked in ichthyology (the study of fishes). In fact, the furthest anyone on the expedition had gotten in analyzing the expedition's fish haul was its leader, C. Hart Merriam, who in an unpublished report referred to the fishes collected in Ash Meadows, including those from Devils Hole, as "Blue Amargosa pumpkin seeds."[29] While this was a kind of naming, this was not a classification; the grouping contained what an ichthyologist would recognize as several different genera of fishes. The expedition needed a professional.

Merriam found his expert in Charles H. Gilbert, who was well placed to make sense of the Death Valley Expedition's collection and to turn jars of "Blue Amargosa pumpkin seeds" and other fishes into distinct species.[30] In 1882, Gilbert had coauthored with his mentor and collaborator, David Starr Jordan, the then-definitive synopsis of North American fish species for the Smithsonian.[31] He had furthered his interest in western fishes while lending his services to the U.S. Fish Commission. On a trip to the Lower Colorado River in 1890, Gilbert even collected pupfish specimens.[32] While Jordan was perhaps the most preeminent ichthyologist of the era, when Gilbert died in 1928, Jordan lauded him as "one of the most careful and accurate of scientific observers...I have ever known."[33] The pupfish were in capable hands.

Gilbert had only recently arrived at Stanford when the fishes from the Death Valley Expedition landed on his desk. Stanford was a new university, opening its doors in 1891, and it had poached Gilbert and Jordan from the Midwest to help establish its campus amid the farms south of San Francisco. Stanford selected Gilbert to head the fishes collection in its zoology department, and Jordan became the university's first president.[34] (While today Gilbert's service to the University is remembered through the Gilbert Biological Sciences Building, Jordan's enthusiastic promotion of eugenics during his lifetime led Stanford in 2020 to remove his name from several campus buildings and spaces.[35]) Stanford's investment in ichthyology and other natural history fields reflected a broader growth in the natural sciences during this era. There was an increasing number of university-run natural history museums, ambitious ocean-going expeditions conducted by the U.S. Fish Commission, and a bloom in scholarly publishing. Looking back on the late nineteenth century, a prominent scientist later called the period a "golden age" for describing fish species.[36]

The findings from the Death Valley Expedition were published in 1893. Crammed into a 400-page tome was information on birds, mammals, reptiles, insects, mollusks, as well as trees and cacti that the expedition had gathered during its months in the field. Gilbert's contribution on ichthyology was also included: it took up just six pages. The short entry concluded that Palmer and his colleagues had collected thirteen species of fishes—including four new to science—despite the expedition's less-than-systematic approach to fish collecting. Two of these new species were from springs in Ash Meadows: the Ash Meadows speckled dace and the Ash Meadows poolfish, which Gilbert named for Merriam (*Empetrichthys merriami*).[37]

But the Devils Hole pupfish was not unique in Gilbert's analysis. In fact, he thought scientists had already described most of the fishes he examined. Gilbert recognized, for example, that the fish from Devils Hole and some other springs in Ash Meadows were part of a genus called *Cyprinodon*. This genus, a guide noted a few years later, was made up of "chubby little fishes, inhabiting the brackish waters of Middle America."[38] Additionally, all of these *Cyprinodon* fishes that the expedition collected, Gilbert wrote, were examples of a particular species, *Cyprinodon macularius*, known to science for over thirty years.[39] Gilbert had a clear reference point: he had already collected this species. Specimens of *Cyprinodon macularius* he had taken on an earlier trip to the Lower Colorado River, Gilbert explained, "have been compared with those from Death Valley and found identical."[40] Far from being unique, the fish in Devils Hole was "identical" to other chubby little fish populations around the West (fig. 1.3).

Yet among these "identical" fish, Gilbert also noticed considerable differences. The "species varies in form and color, and apparently the size which it reaches in different localities," he observed.[41] Considering the fifty-eight specimens of *Cyprinodon macularius* he had from across Ash Meadows and the Amaragosa—habitats he had never seen—Gilbert noted that some of the fish had unique coloration, that some had a black bar across the end of their caudal (tail) fin, and that the number of fin rays (the spiny structures supporting the fins of fishes) could vary from fish to fish and population to population. He even singled out the Devils Hole specimens as particularly distinct, after observing that all ten of the small specimens were missing pelvic fins, suggesting that "further collections from this locality would be of interest."[42]

Today, we define the fishes Gilbert grouped under *Cyprinodon macularius* as belonging to several species and subspecies. And it is the uniqueness of the Devils Hole pupfish, in particular, that has motivated several genera-

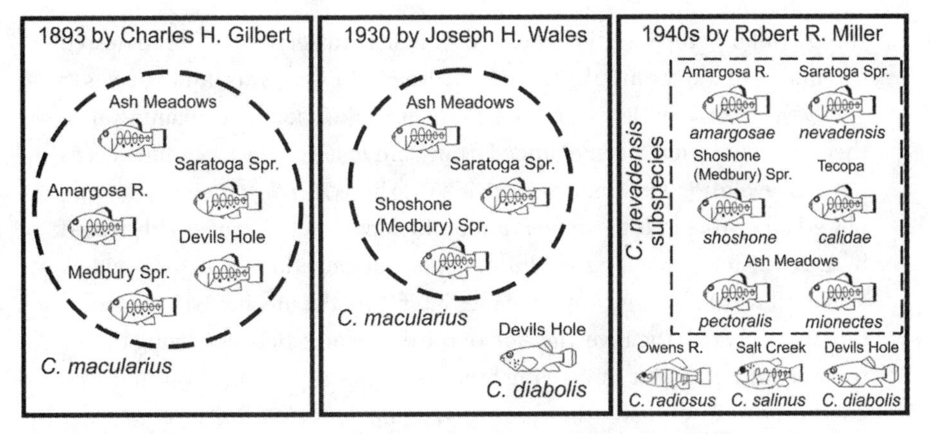

FIGURE 1.3. Classification of genus *Cyprinodon* in the Death Valley region from the 1890s through the 1940s. The Tecopa pupfish (*C. nevadensis calidae*) is now extinct. An additional subspecies of pupfish (*C. salinus milleri*) was later discovered at Cottonball Marsh in Death Valley. Art adapted from *A Progress Report on the Status of the Desert Pupfish*, U.S. Department of the Interior, June 1971.

tions of scientists, managers, and conservationists to ensure its survival. If it were still considered "just" a population of a wide-ranging species that occurs in springs across the Southwest, it is doubtful the pupfish in Devils Hole would have been so lucky.

The survival of the pupfish hinged on its distinct classification, but in its first brush with science and the federal government it was not seen as special. Why did Gilbert not describe the Devils Hole pupfish as a unique species? And why was it later reclassified? The answers have to do with the purpose of species classification in late-nineteenth- and early-twentieth-century taxonomy and even more so with the specific laboratory methods Gilbert and other ichthyologists used to analyze specimens.

As noted earlier, humans are profligate classifiers and namers. We cannot help ourselves. Scientists, especially, have succumbed to this impulse. The task of classifying plant and animal life goes back to Aristotle, but its modern origins came in the mid-1700s when a Swedish naturalist named Karl von Linné (better known as Carolus Linnaeus) proposed a Latin-based naming and classification system that is still in use today despite substantial changes in form and meaning. Generations of middle school students can, in part, blame Linnaeus for being forced to learn the mnemonic device

"King Phillip Came Over From Greece Singing" to keep track of the nested categories of kingdom, phylum, class, order, family, genus, and species.[43]

By the time scientists stumbled on Devils Hole in 1891, the meaning of all these categories had been changed from Linnaeus's era by the rapid acceptance of evolutionism. Charles Darwin's 1859 book, *On the Origins of Species*, played a key role in the propagation of the idea, but the process identified by Darwin for propelling evolution—natural selection—was not widely accepted within biology until decades later.[44] As the mechanism for evolution was debated, the investigation of the differences between living things became important for unraveling the mysteries surrounding how and why animals, plants, and all other life evolved and were distributed across the globe. In the late nineteenth century, instead of just answering "what" questions (what kind of animal is that?), naming and classifying species could unravel "why" questions about the origins and diversity of life on Earth.

One challenge for natural scientists was to use the evidence available—the bodies of animals and plants—to devise classifications that reflected evolutionary relatedness.[45] (This practice of fitting taxonomic classification and naming into evolutionary relationships is today generally referred to as systematics, or systematic biology.) Especially difficult for taxonomists was the job of discriminating between similar populations and deciding whether they constituted separate species, then examining the proposed species to see whether it was itself composed of many unique populations that deserved the label of subspecies.

This returns us to the jars of fishes at the Smithsonian. Classification started in the field with collecting, but answering questions about whether a group of similar fishes was many species or just one depended on work done in the laboratory carefully analyzing the collected material. For ichthyologists, it meant measuring the shape (morphometrics) and counting features (meristics) of fish under a microscope. Gilbert, for example, counted the number of rays on each fin, noted the proportion of the head relative to the rest of the body, measured the size of the eye socket, tallied the number of scales between a fish's gills and caudal (tail) fin, and observed the relative positioning of fins along the body. This was meticulous, difficult work—especially with fish less than an inch long—and required keeping track of tiny variations between individuals.

In analyzing the Death Valley fishes, Gilbert "lumped" several geographically isolated populations under the heading of one species, despite observing and recording differences in the fish. He did so, in part, because the Death Valley Expedition came back with relatively few individual fish, and he could not say whether the variations he observed were characteristic of all the fishes in these habitats or peculiarities of the individuals he had the opportunity to examine. Any one fish could be quite different, so ichthyologists preferred to study many specimens to tease out how populations (not just individuals) differ from one another statistically. In noting these specimens' difference, Gilbert was further hamstrung in the separations he could make by the fact that he had not been to Devils Hole or any of the other habitats. He even assumed that all the pupfish Palmer collected at Devils Hole were juveniles because of their diminutive size relative to other populations.

Gilbert's decision to group all the populations together may also have reflected his personal preferences. In 1884, Elliot Coues, a prominent ornithologist declared, "No infallible rule can be laid down for determining what shall be held to be a species...It is a matter of tact and experience."[46] Ichthyologists accepted Coues's ambiguity—David Starr Jordan's classic ichthyology textbook quoted this passage from Coues as he discussed classification.[47] Taxonomists might refer to themselves or others as "lumpers" or "splitters." The former would allow for more variation across multiple populations and keep them in one species, while a splitter would take even small variations between populations as a reason for naming a new species or subspecies. Gilbert lumped.

By 1893, based on the limited extent of the specimens collected by the Death Valley Expedition, and the interpretation of a respected ichthyologist, the state of knowledge suggested that one species of pupfish occurred across Nevada, California, and Arizona: *Cyprinodon macularius.* The pupfish in Devils Hole had a name, but it was not unique; it was just a population of a species that lived in desert habitats across the Southwest.

• Isolated Character

"If fishes were as easily preserved, measured and examined as birds, and if they were studied by as many keen eyes," David Starr Jordan lamented in 1905, ichthyologists might realize that many existing species could be "broken up" into other species and subspecies. "This is certainly probable in the case of fresh-water fishes," he said.[48]

Jordan may as well have been speaking directly about the desert fishes uncovered by the Death Valley Expedition. While the party had multiple members competent with identifying mammals and birds, it had no specialist in the field with them who focused on ichthyology. And although there were few eyes keener than Charles Gilbert, with a small sample size he could only lump Death Valley *Cyprinodon* into one species. There was still work to be done to unravel a classification system that captured the evolutionary relationships among fishes in Death Valley.

Revisions of previous classifications were part of this scientific practice. In 1896, in the preface to an update of his 1882 volume on North American fishes coauthored with Gilbert, Jordan commented that "in many of its details" his system of classification was "purely tentative, to be confirmed or changed when the anatomy of the various forms [populations] is better known."[49] With additional collections, and more careful scrutiny, the status of what was known to be a species, or a genus, or a family, and so on, was expected to shift. This openness to reexamination is also one of the reasons that all those jars of fishes remain on shelves at the Smithsonian and in natural history museums around the world: they are not one-time use but available—like library books—for repeated study and reinterpretation.

The first hint that the pupfish might get a second look came more than thirty years after the Death Valley Expedition, when a young curator at the University of Michigan Museum of Zoology named Carl L. Hubbs started investigating the way that ichthyologists had classified the genus *Cyprinodon* and announced in 1926 that he was "preparing a revision of the numerous species of the genus."[50]

One of the most prominent ichthyologists of the twentieth century, Hubbs's career stretched from the 1910s to the 1970s. Trained at Stanford under the tutelage of Jordan and Gilbert, Hubbs left the Bay Area after earning his master's degree and became a curator at the Field Museum of Natural History in Chicago for three years before being hired at the University of Michigan. There, he was later awarded a doctorate on the basis of his prolific publication record. In Ann Arbor, Hubbs built the university's fishes collection from about 5,000 to more than 2 million specimens, with an especially pronounced focus on freshwater fishes. Later, in the mid-1940s, he departed Michigan and returned to the West Coast for a senior position at the University of California's Scripps Institution of Oceanography, in La Jolla.[51]

Respected for work that explored everything from the Great Lakes to the Pacific Ocean, Hubbs was also famous for his unstoppable energy. Eugenie Clark, in her classic account of mid-twentieth-century oceanography, *Lady with a Spear* (1951), reported meeting Hubbs at a party during an ichthyology conference in Pittsburgh. She described him as "the life of the party" but later discovered that he "was as serious at his work as he was jovial at his play."[52] Moving to California to study with Hubbs, Clark arrived in La Jolla dead tired and ready for sleep after a manic, nearly nonstop drive from New York, only to be immediately pressed into assisting Hubbs and his wife, Laura, with a midnight inventory of grunions spawning on the beach.[53] Other graduate students also fell into Hubbs's frenetic orbit. Ken Norris, later a well-known ecologist and a founder of the University of California's extensive system of biological reserves, recalled how automobiles were an outlet for that "Hubbsian energy." Hubbs, Norris remembered, once traversed a rutted dirt road in Baja California at forty miles per hour, while eating his lunch and sharing a lecture on the surrounding desert with him. "Wherever fish are, Carl Hubbs goes," said Norris.[54]

When a correspondent told Hubbs he wished he could live for two or three hundred years so that he could see the fruits of his work in conservation borne out, Hubbs replied that he would rather he could live two or three lives simultaneously. There was so much work to do right now, both in addressing scientific problems and contributing to the conservation movement. "I feel constantly remorseful for the small amount of time and effort I can put into work toward the preservation of some of our natural conditions and wild life, both above and below the water line," he wrote.[55]

The reality that Hubbs lived just one life caught up with him when it came to refining the classification for pupfishes. Despite his 1926 pronouncement, more than a decade later Hubbs could still write, "My promised revision of *Cyprinodon* is one of a hundred projects underway but not published."[56] He had not collected fishes in the Death Valley region, and, given that one of the central limitations of Gilbert's work was the limited availability of specimens, this also did not bode well for a revision. Hubbs ultimately published little on the pupfish of the Death Valley region, but he became a center of gravity for both pupfish science and, as we will see in the next chapter, pupfish conservation.

While Hubbs focused his energy elsewhere, two Stanford graduate students traveled to Ash Meadows in the spring of 1930—the first scientific foray to Ash Meadows since the Death Valley Expedition thirty-nine years before. Getting to Devils Hole was substantially easier than it had been for Palmer and his crew: Joseph H. Wales and his friend, George Myers, relied on a Ford Model A instead of horses and carts. Years later, Wales recalled that they were basically stressed out at school and wanted a few days away from campus, so they set off for some collecting in the desert. A local rancher pointed them to Devils Hole.[57] With a seine, Wales and Myers also collected fishes Gilbert had identified as *Cyprinodon macularius* at three other springs in Ash Meadows and two more southward along the Amargosa River. Crucially, they collected many more specimens at each location. At Devils Hole, instead of the ten fish taken by Palmer in 1891, Wales and Myers took sixty. At King's Spring, part of the Point of Rocks group of springs and seeps two miles southeast of Devils Hole, they collected two hundred.

Returning to the laboratory at Stanford with this much larger haul of evidence, Wales set about examining the rays on pectoral, dorsal, anal, and pelvic fins, as well as the number of scales running along the side of the body, the number of vertebrae, and each specimen's length. After this examination—more thorough than Gilbert's—Wales discovered that each population had a range of characteristics. While some of them remained very similar across the isolated populations, others varied considerably both within populations and between them. The average number of rays on the pectoral fins of the population from King's Spring, for example, was 15.5, while the Devils Hole population had an average of 16.32 rays. The most distinctive feature of the Devils Hole pupfish was its pelvic fins: only one specimen even had these fins.

Wales could spot differences in the average of these measurements, but he did not consider them different species. "No character in any population," he noted, "failed to overlap the corresponding character in the other populations. . . . So far as these characters are concerned, therefore, we must say that all of our fish belong to one species."[58] In other words, with each measurement he took, variability within one population overlapped with another. The Devils Hole and King's Spring populations had different mean numbers of pectoral fin rays, but their ranges overlapped: the Devils Hole population had between 15 and 18, while the King's Spring population had between 13 and 17. Even the most identifying feature of the Devils Hole pupfish, the absence of pelvic fins, Wales found in other populations (though much less frequently). If all the fish he collected were dumped on a table and mixed together, Wales's analysis would make it difficult, if not impossible, to sort out which fish came from which pool. "No distinction," Wales wrote, "can be made between any of the groups in respect to the characters which were studied statistically."[59]

Despite the limitations of his analysis, Wales decided that the pupfish from Devils Hole was indeed its own species. While difficult to capture through statistical measurements, the Devils Hole pupfish was still recognizable, he thought, by its bluish hue and small size. Furthermore, the "isolated character" of the Devils Hole habitat meant that it was "quite probable that there has been no intermingling of these fish with those outside for a much longer period than the period of isolation for any of the other groups."[60]

Still, it was possible, he said, that future, controlled experiments would show that the unique appearance of the Devils Hole pupfish had less to do with evolution and more to do with the high temperature of the water in Devils Hole changing the appearance of the fish. Raising Devils Hole pupfish under other conditions and seeing whether the fish developed differently would help resolve the issue. But until such an experiment proved it "merely a variation of *C. macularius*," Wales thought it best to "recognize the differences by a name and description."[61]

A host of questions remained that Wales's paper could not answer. How different was the Devils Hole pupfish from other populations in Ash Meadows and the region? Was its diminutive size a result of its genes or its extreme environment? Or both? And would further collection from other springs—and their close examination—reveal additional variation between pupfish populations? Despite such uncertainty, Wales's paper transformed pupfish in Devils Hole into the Devils Hole pupfish, *Cyprinodon diabolis*.

Carl Hubbs learned about Wales's paper right after it had been drafted in June 1930. Hubbs was then the editor for ichthyology at the journal *Copeia*, the shared scholarly home for ichthyology and herpetology (the study of amphibians and reptiles) in the United States, and Wales had submitted the paper there for publication. In Hubbs's review of the manuscript, which he accepted to the journal without need of revision, Hubbs told Wales he considered the "*Cyprinodon* problem" to be "fine meat" for a doctoral dissertation. "Please have no regard for my toes if you wish to go on with such work," Hubbs wrote.[62] Wales, however, pursued a master's thesis on blue rockfish and later worked for California's wildlife agency and then Oregon State University.[63] He never published again on pupfish.

Although Wales left the pupfish behind, his paper—and the specimens he collected—lived on. He could not have known it at the time, but the article represented a dramatic moment for how scientists, conservationists, and the general public have thought about the pupfish. From Wales's decision to separate the Devils Hole population as its own species, people came to view the pupfish differently. And this taxonomic choice had profound implications for wildlife conservation, water and land use, politics, and economics in southern Nevada. As we will see, the uniqueness of the pupfish served as the basis for the addition of Devils Hole into Death Valley National Monument, a lengthy battle over water rights, the species' position on the endangered species list, and also played a central role in the pupfish's persistence into the twenty-first century. Naming a species had consequences.

• Pupfish Teeth

In spring 1937—almost seven years after Wales's suggestive paper on the Devils Hole pupfish—Carl Hubbs received word from a colleague at the California Division of Fish and Game that a student at the University of California, Berkeley, had been traveling around the desert collecting fishes—and had even made a trip to Devils Hole. Hubbs vowed "to find out just what the lad is up to."[64]

After making an inquiry to the University of California Museum of Vertebrate Zoology at Berkeley whether there was a student meeting this description, Hubbs received a letter from an undergraduate named Robert Rush Miller. A Los Angeles native who turned twenty-one in 1937, Miller still had two years of coursework to complete before receiving his bachelor's degree. Encouraged from a young age in his interest in natural history by his father, Miller had been exploring the California backcountry for years. During a geology class field trip at Pomona College, which he attended before transferring to Berkeley, Miller first saw desert fishes in the Mojave River and was enthralled.[65]

Starting in 1936, with his father as his field partner and assistant, Miller began more ambitious field collecting efforts. In his letter to Hubbs, Miller explained he had collected pupfish from the Death Valley region and brought them back to Berkeley alive, writing that "they seem to be doing excellently." Miller had even visited Devils Hole—after Palmer and Wales, perhaps just the third scientist to do so; although he revealed his novice status when he incorrectly identified the fish living there as the Ash Meadows poolfish (*Empetrichthys merriami*), rather than as the Devils Hole pupfish.

Whatever he lacked in training, Miller made up in enthusiasm, telling Hubbs of his plans to collect pupfish across the Southwest that summer: "I have become profoundly interested in the study of fish, especially the systematics of California fishes, and the more I delve into the subject the deeper my desire for knowledge becomes."[66] Miller's letter was music to Hubbs's ears. While science is driven through questions, methods, and theories developed by a community and is enabled through institutions like museums and universities, it also can depend on the small connections that bring particular researchers together. Wales had declined to pursue further research on desert fishes, and Hubbs still was having trouble living two or three lives at once. Perhaps Miller would take on further investigations of the *Cyprinodon* genus?

Miller and Hubbs struck up a correspondence, with Miller periodically updating Hubbs on his studies and collecting efforts. The following summer, in 1938, Hubbs invited the student to join him and his family for a few days on an extended collecting trip through the West. Hubbs often enlisted not only his wife in his collecting expeditions but also his three children, who served as field assistants. The children earned allowances for their efforts, including bonuses for collecting new species and genera. (As two of Hubbs's children later recalled, their pocketbooks grew quickly since Hubbs was a "splitter.")[67]

Miller was thrilled. He had been invited to join one of the most prominent ichthyologists of the day on a trip through the desert basins of California and Nevada. As arranged, Miller and Hubbs met in person for the first time on July 5, 1938, at Lehman Caves, Nevada, now part of Great Basin National Park in northeastern Nevada. In his field notes, Miller wrote of Hubbs on meeting that he was "stocky, robust and full of energy."[68] As with the Death Valley Expedition and Wales's trip to Ash Meadows, the purpose of this fieldwork was to collect and preserve specimens that could be examined back in the laboratory to tease out the relationships between the fish populations of the Great Basin.

Miller immediately hit it off with Hubbs and his family. And Hubbs apparently recognized Miller's potential. Near the end of the ten days that Miller had planned to join them, Hubbs asked Miller if he would like to stay on with his family's expedition for another few weeks, with Hubbs covering his expenses. "I was so dumbfounded at the time," Miller wrote in his field notes, "I could gather no words for an answer but I didn't get to sleep very quickly tonight what for thinking about this generous offer and marvelous opportunity!"[69]

Miller also spent the summer of 1938 falling for Hubbs's daughter, Frances. The Hubbs family dropped him off at a Nevada bus station for the journey back to his final year at Berkeley, and within hours he was mailing Fran a letter.[70] Miller was smitten. The following summer, Miller moved to Michigan, where he became a graduate student under Hubbs. He also married Fran, making him both son-in-law and advisee to Hubbs.

While at Michigan, Miller took up an examination of the *Cyprinodon* around Death Valley with gusto. From 1939, when he arrived, until 1944, when he finished his doctoral dissertation, Miller worked on little else. The resulting dissertation was dramatically more ambitious than anything previously published on desert fishes and still serves as an important reference for scientists and managers. Miller revisited not just Devils Hole but every other spring he could find in the Death Valley region. In total, between 1936 and 1942, he personally (frequently with his father or his wife at his side) collected more than 10,000 specimens of *Cyprinodon* from California and Nevada. While "overcollecting" can have some potentially disastrous consequences, discussed in the next chapter, scientifically these specimens were incredibly valuable. Previous studies of the fishes of the Death Valley region, besides Wales's look at Devils Hole, had been limited "principally because adequate series [of specimens] have not been available."[71] With this extensive fieldwork, Miller's analysis would not suffer from the same problem.

Miller returned to the laboratory in Ann Arbor and devoted himself to the careful counting and measuring of pupfishes. Whereas Wales had made 8 major measurements on a total of 650 fish from across the 5 habitats he examined, a project that could have taken him no more than 2 months (the time between his collection and his submission of the completed manuscript to *Copeia*), Miller examined more than 25 different features on each fish and did so for thousands of pupfish specimens from more than 20 locations in Ash Meadows, Death Valley, and the Owens Valley.

Miller often counted or measured the same feature multiple times to ensure he was accurate. All of this, he wrote—with serious understatement—required "considerable practice." He did not even try to examine the Devils Hole pupfish until he was proficient with comparatively larger specimens from other springs. At an average length of just 19 mm, not counting the caudal fin, some features were very difficult to evaluate, even under a strong microscope and with precision calipers. "It was... difficult to measure" some of the bones around the Devils Hole pupfish's eyes, Miller wrote, since the distance was under one millimeter.[72] In his files at the University of Michigan Museum of Zoology, I came across pages and pages of data sheets with measurements. If the pupfish's survival across the twentieth century is an act of persistence, then Robert Miller's detailed work in identifying the variation within Death Valley-area pupfish populations is also a noteworthy study in human persistence.

From this painstaking examination, Miller found minute but substantial variation, becoming someone who could use the teeth of pupfish to tell them apart. He could show that pupfish from the region had identifiable differences in scale structure and size, tricuspid tooth shape, length, body shape, color, position of dorsal fins, and other features. Especially when multiple attributes were considered simultaneously, the pupfish from across Death Valley looked remarkably different. Instead of a sea of *Cyprinodon macularius* that Gilbert found from the Death Valley Expedition specimens, Miller's results showed coherent islands of variation.

Based on those measurable differences among populations, Miller regrouped the Death Valley *Cyprinodon* into four different species, with one of these species, *Cyprinodon nevadensis*, made up of six distinct subspecies (fig. 1.3). Subspecies, in the view of Miller's advisor, Carl Hubbs, were "animal kinds . . . sufficiently clear-cut as to be thought worthy of a place in the nomenclatorial system, but which do not give evidence of being completely differentiated."[73] A subspecies classification, then, was meant to capture subtle evidence of evolution at work: isolated in different aquatic habitats, natural selection had begun sending populations of *C. nevadensis* in different evolutionary directions. It was exactly what Miller hoped to find evidence of through his years of focus in the laboratory.

Miller's careful work also confirmed what Joseph Wales, in Miller's words, had "provisionally described": the Devils Hole pupfish was a unique species. Its "dwarfed" size, lack of pelvic fins, and several other features confirmed its "specific distinctiveness," Miller wrote. After all his attention to the specimens, Miller even concluded that *Cyprinodon diabolis* was the "most unusual species in the genus."[74] In Miller's careful analysis, the Devils Hole pupfish was even unique among unique pupfish.

Miller also undertook experimental work to see how much of the distinctiveness of different pupfish populations was genetic or environmental. This effort went toward unraveling the question Wales posed in his 1930 paper: Was it the characteristics of the pupfish themselves (their genes) or the conditions in which they lived that made them appear so different? His results on this front were less conclusive.

Miller conducted hybridization experiments with pupfishes in small, artificial ponds his father maintained in his Los Angeles backyard, though never with the Devils Hole pupfish itself. Getting the fishes alive back to LA was a chore, but between 1940 and 1944, the elder Miller raised several generations of pupfish and hybridized them in different combinations to see how their morphology changed. (Miller also transplanted pupfish to natural springs across the southern deserts of California that did not have any fish life at all. He did not publish anything about these transplants until the late 1960s, mostly to alert future researchers as to why they might find unexpected species in particular bodies of water.[75])

Neither of these approaches gave Miller a clear answer as to the influence of genes and environment on the pupfish, but he believed his data "strongly indicate[d] that many of the characters are wholly or in part genetic."[76] Later, discussing the Devils Hole pupfish, he told a correspondent that he "always wanted very much to experiment with this bizarre species." In particular, Miller wondered "whether raising the species at a considerably lower temperature than its natural habitat would result in the reappearance of pelvic fins."[77]

This question, left hanging in the 1940s, reemerged in pupfish research in the 2000s, when scientists conducted an experiment showing that genetic uniqueness may only partly explain the Devils Hole pupfish's appearance. By raising the related Amargosa River pupfish in a laboratory with temperature and food availability conditions similar to Devils Hole, the fish grew to look more (though not exactly) like the Devils Hole pupfish: larger heads, smaller body size, and reduced pelvic fin development. As a group, these pupfishes exhibit "phenotypic plasticity," with their appearance also influenced by the conditions under which they are raised, not just their genetic code.[78] The apparent differences in pupfish morphology—so central to the consequential classification of the pupfishes of the Death Valley region during the 1930s and 1940s—were, at least partly, a result of different environmental conditions, not different genes.

In addition to Miller's work in elaborating a classification system based on an extensive examination of pupfish morphology and some limited experimentation, he also incorporated a historical-geological analysis to explain his groupings. Basically, Miller tried to answer the question: how would fishes get into the desert in the first place? In a series of papers, Miller, along with Hubbs, investigated how during the Pleistocene (the epoch now considered to stretch from 2.5 million years to 12,000 years before the present) many of the desert basins in what they called the "Death Valley system" had been connected through a changing series of lakes and rivers. The "pluvial," wetter period in the history of Death Valley would have enabled an ancestral species of pupfish to live across this very different landscape.

They then reconstructed a scenario for the drainage of the late Pleistocene lakes and rivers in the region, showing that the fish species could have become isolated in their current habitats as the lakes and rivers dried up. The pupfish in Devils Hole, the highest elevation spring in Ash Meadows, was "very likely [to] have been isolated from the adjacent populations...in Ash Meadows for a much longer period than have any populations throughout the Amargosa River basin," wrote Miller. "As the waters subsided, a stock [of ancestral pupfish] became isolated in Devils Hole and has been effectively cut off from communication with other stocks ever since."[79] The Devils Hole pupfish looked different and qualified as a unique species because it had been cut off from the rest of the pupfishes for a long time, presumably since before the end of the Pleistocene. Evolution had worked on the Devils Hole pupfish as it persisted and changed in isolation for thousands of years.

Today, that the Devils Hole pupfish is a unique species seems self-evident. Yet defining the population in this way was a process forged, not only through evolution, but also through the development of a particular method for studying and classifying fishes over the course of more than forty years. This is not to say that the pupfish is not "real" or unique—it clearly is in the ways that Miller described. But when we look at how it has persisted, we must acknowledge the role played by a particular scientific practice that culminated in decisions to call it one thing and not another.

Joseph Wales himself understood this well. In a 1985 letter, he reminisced about his 1930 visit to Ash Meadows and Devils Hole. Wales considered his reflection, mailed to a conservation organization focused on desert fishes, confidential. In it, he questioned his own judgment in declaring the Devils Hole pupfish to be a unique species. "If I were to do it all over again," Wales wrote, he might have named the Devils Hole population a subspecies, or not given it a new classification at all. "Obviously in 1930 I was in the 'splitter' stage of my development," he wrote.

Wales had apparently forgotten that in the decade after his paper appeared Robert Miller's work showed, through his much more detailed laboratory analysis, that the Devils Hole pupfish was certainly worthy of distinct species status. Yet Wales knew what it would mean to walk things back now. By the 1980s, a surprisingly long list of people could have used a taxonomic revision as ammunition for calling into question the extensive protection the species had received as a result of its uniqueness. "For reasons of conservation we should retain the above name [*Cyprinodon diabolis*]," he wrote. "Conserving a unique species is, in the minds of most people, a more noble deed than conserving a race or variety."[80]

Indeed, the results of Wales and Miller's classifying work, as much as the pupfish's evolutionary history, paved the way for its persistence, as advocates translated taxonomic understanding into arguments for its protection. In 1952, President Harry S. Truman added the 40 acres around Devils Hole to Death Valley National Monument. The central rationale for this action—as we will see in the next chapter—was that the Devils Hole pupfish, according to Truman's proclamation, was a "peculiar race of desert fish...which is found nowhere else in the world, evolved only after the gradual drying up of the Death Valley Lake System isolated this fish population from the original ancestral stock that in Pleistocene times was common to the entire region."[81]

Robert Miller or Carl Hubbs could not have said it any better. In fact, Truman's proclamation echoed their research and resulted from their activism.

CHAPTER 2

To Protect and Conserve[1]

During the same month in 1891 that T. S. Palmer visited Devils Hole, he also gathered specimens from King's Spring, just two miles away. Along with the other fishes collected by Palmer and his colleagues, Stanford's Charles Gilbert described them in the Death Valley Expedition report. Unlike the pupfish Gilbert analyzed, however, he recognized the King's Spring specimens as members of a species new to science. As we have seen, he called the species *Empetrichthys merriami*, later dubbed the Ash Meadows poolfish.[2]

Also in contrast to pupfishes, which were collected prodigiously in Ash Meadows during the 1930s and 1940s, few specimens of the poolfish were added to the Death Valley Expedition's initial tally. Robert R. Miller captured just twenty-two more poolfish from his many visits to Ash Meadows despite having "made special effort to obtain greater numbers."[3] As late as 1967, Miller held out hope that the species still existed, but no one observed or collected the species after 1948.[4] The fish was never seen again; the Ash Meadows poolfish is extinct.[5]

The Ash Meadows poolfish became wrapped up in the biological sciences as the result of the same foray into the desert as the Devils Hole pupfish, but in this case a unique classification and name recorded in an official 1893 report did not protect the species. In other words, the Ash Meadows poolfish was not saved through its engagement with government or the scientific community. While classification and naming living things helps make the Earth's biodiversity comprehensible, the process does not automatically result in conservation. The case of the poolfish is a reminder that although classification plays a necessary role in survival for many species, it is not sufficient.

Figuring out exactly what is "sufficient" to shield a species from extinction is easier said than done. For one thing, protecting species—enabling survival—does not only result from a single act in a particular moment but also from a continual, perhaps unending, relationship between a species—including its life history, habitat, and behavior—and human institutions.

FIGURE 2.1. Devils Hole in 1947, by Nevada artist Gus Bundy. Courtesy Gus Bundy Collection, UNRS-P1985-08-00280, Special Collections and University Archives Department, University of Nevada, Reno.

For another, and as the pupfish story clearly shows, it is not always easy to know what kinds of measures may be best suited to enable survival.

During the 1940s and 1950s, as the poolfish slid toward extinction, scientists and government agencies wrestled for the first time with these issues, discussing whether and how to protect the Ash Meadows landscape and its unique animals. It did not begin well. When Carl Hubbs first approached the National Park Service about the idea of protecting Devils Hole and other Ash Meadows springs in 1943, administrators, including the agency's future director, Conrad Wirth, dismissed the idea. To Wirth, the whole concept of protecting pupfish "seem[ed] to be purely a scientific matter," not a responsibility of the Park Service.[6] This, from the agency that eventually *did* protect Devils Hole.

Other developments during that decade were also disheartening. The Park Service's initial thanks-but-no-thanks was actually better than a now long-forgotten Fish and Wildlife Service (FWS) plan for the area. Buried in the National Archives in Maryland is a 1947 FWS proposal that would have radically reshaped Ash Meadows' aquatic habitats, likely threatening a number of endemic species directly. And starting the previous year, in what can best be described as conservation vigilantism, a University of Nevada researcher took actions that, despite good intentions, seriously threatened the Devils Hole pupfish.

This chapter explores how the Devils Hole pupfish survived such mishaps and became included in Death Valley National Monument—the first formal step taken to protect it from extinction. It is important to review this part of the pupfish's history, in part, because of how inadequate the addition to the monument later turned out to be for ensuring the pupfish's persistence, setting the stage for future conflicts over the pupfish and the best uses for the desert.

The story also unravels a simple good guy/bad guy dualism common in conservation stories and one especially tempting for telling the story of the pupfish, where ranchers and real estate developers in Ash Meadows later played the villain. Instead, the early efforts to protect the pupfish force us to hold, perhaps uncomfortably, two contradictory ideas in our heads at once: that the Devils Hole pupfish persisted both despite *and* because of actions by institutions and people charged—by law or love—with protecting it. Armed with this approach, the events that brought the pupfish into the national park system in 1952 look both more incredible and more incomplete.

• It Won't Be Long

Climate change has made species extinction a prominent moral and political issue. It is often listed, along with rising sea levels and drought, as a main ecological outcome of a warming planet. But extinction is not new, and neither is concern over it. As historian Mark Barrow has shown in his masterful 2009 book on the history of extinction, *Nature's Ghosts*, scientists discovered the concept in the early nineteenth century and have been "increasingly haunted" by it ever since.[7] When Robert Miller, Carl Hubbs, and others came to Ash Meadows, they carried baggage with them: their concern for the fishes they studied stemmed from a broader fear of extinction as well as emerging ideas about how best to prevent it. Essentially, the Devils Hole pupfish benefited from past extinctions.

Some nineteenth-century extinctions resulted from direct human persecution. In the U.S., the extinction of the passenger pigeon and a near miss with bison are among the most widely known. In both cases, unregulated subsistence, commercial, and sport hunting featured prominently.[8] But the bison and the passenger pigeon were not alone. As William T. Hornaday, a taxidermist who underwent a near-spiritual conversion and played a critical role in rescuing the bison from extinction, commented in his 1913 book, *Our Vanishing Wildlife*, "There is not a single state in our country from which the killable game is not being rapidly and persistently shot to death, legally or illegally, very much more rapidly than it is breeding, with extermination for the most of it close in sight."[9]

A second reason for extinctions was also cause for alarm by the early twentieth century. F. B. Sumner, a prominent ecologist at the University of California's Scripps Institution of Oceanography in La Jolla—where Carl Hubbs later worked—cautioned his fellow scientists not to focus "attention too exclusively upon these relatively few examples which are so conspicuous—the mammals and birds which are sought for as sources of food or feather or fur." Hornaday, in other words, only captured part of the problem. What Sumner found as disturbing was the pace of industrial development that could transform entire ecosystems and result in species being lost in the shuffle. "Forests are vanishing, brush land being cleared, swamps drained and the desert irrigated," he wrote.[10] Species were being harmed by altering the habitats where animals lived, not just through hunting.

Aquatic species in the arid U.S. West were especially vulnerable to habitat loss, since rare water sources often served as focal points for a range of human activities. Harvard University biologist Charles Brues, for example, wrote to his colleagues in 1928 that "all except the most inaccessible" springs in the West "have been converted into natatoria, sanatoria for arthritics, radium baths and the like, or have been diverted into irrigation ditches, sometimes with the aid of dynamite, to supply a few desolate ranches with water for cattle and alfalfa." The outcome of this process was grim, warned Brues: "The fauna and the flora of these thermal springs is destined to be wiped out over considerable areas by the advance of commercialism."[11]

Fishes bore the brunt of such changes. Recent research suggests that freshwater fishes had the highest extinction rate of all vertebrate animals during the twentieth century. In North America alone, between 1900 and 1950, some 14 species or subspecies of freshwater fishes became extinct. The rate accelerated between 1951 and 2010, with at least an additional 43 extinctions. Of these 57 total taxa, eight were pupfishes.[12] Ichthyologists did not undertake a comprehensive list of extinct fishes until after World War II, but even before then scientists knew fishes were in trouble. Robert Miller responded to the news that a pupfish in Texas was thought to be extinct following dam construction in its habitat, writing, "Extinction of *Cyprinodon* in isolated springs... seems to be of fairly common occurrence as far as my observations in California have gone."[13] For pupfishes, then, the problem was not that they were being sought out for meat pies (passenger pigeon) or robes (bison) but that their habitats were vulnerable to a range of practices from ranching and irrigated agriculture to the development of municipal water supplies and urban expansion.

Scientists, including Miller, brought this foreboding over the prospects for isolated species with them to Devils Hole. In 1937, returning from his second trip to Ash Meadows in six months, Miller reported some alarming findings. Compared to his first visit, he believed the population of Devils Hole pupfish was sharply reduced. In a letter to his future graduate advisor (and father-in-law), Carl Hubbs, Miller estimated the entire Devils Hole pupfish population at fewer than fifty individual fish and declared, "At this rate it won't be long until they are extinct."[14] After repeating his discovery to George S. Myers, who had accompanied Joseph Wales on his 1930 trip to Devils Hole, Myers replied to Miller that his "memo concerning the trials of *Cyprinodon diabolis*... is sad, but to be expected, I am afraid."[15] Later, in 1938, Wales, the describer of the Devils Hole pupfish, wrote to Miller and wondered somewhat idly whether "the population still exists."[16]

Miller's correspondence on the extinction risk facing the Devils Hole pupfish reflects the broad concerns for isolated species among natural scientists in the 1930s. It is not, however, an especially useful guide for the actual risks facing the Devils Hole pupfish. While Miller later became a leading expert on desert fishes, when he raised the alarm over the Devils Hole pupfish in 1937 he was an undergraduate student with little expertise. Just six months before he predicted the Devils Hole pupfish's extinction, Miller did not even know the correct species in Devils Hole. In a letter to Hubbs he had misidentified it as *Empetrichthys merriami*, the Ash Meadows poolfish (which is ironic, given the fact that the poolfish *did* later become extinct).[17]

Miller's anxiety over Devils Hole channeled his chosen discipline's broader unease, not anything specifically amiss in Devils Hole.

Natural scientists responded to concerns over extinction in two key ways: through their regular scientific practice and by making a novel argument for habitat protection. First, extinction became an important rationale for one of their main methods: collecting, cataloguing, and storing specimens. Such "museum work," historian Peter Alagona has written, constituted "a form of conservation."[18] While this practice long predated the early twentieth century, as did natural history museums themselves, both underwent expansion and methodological standardization during this era.[19] One of the leaders in this movement of early-twentieth-century natural history museums was Joseph Grinnell, founding director of UC-Berkeley's Museum of Vertebrate Zoology. Grinnell completed his graduate work at Stanford University partly under the direction of a familiar character in the pupfish story: Charles H. Gilbert, the first scientist to examine pupfish from Devils Hole.[20]

Beginning in 1908 with the museum's establishment, Grinnell and his colleagues scoured California, collecting and preserving animals from nearly all corners of the state, ultimately describing, mapping, and preserving a thorough record of its fauna. "Many species of vertebrate animals are disappearing; some are gone already," Grinnell wrote. "All that the investigator of the future will have...will be the remains of these species preserved more or less faithfully, along with the data accompanying them, in the museums of the country."[21] He could not prevent overhunting or habitat loss from logging or clearing for agriculture, but Grinnell and museum curators around the country could conserve records of when and where certain animals lived and store specimens for subsequent generations.

On its own, storing the remains of animals was a rather narrow—and fatalistic—vision of conservation. Natural scientists, however, many connected with a fledgling professional organization founded in 1915 called the Ecological Society of America, also made a second, broader mark on conservation by advocating for the protection of entire habitats and ecosystems.[22] Through the purchase of private lands or designating public lands for wildlife, such natural areas could serve not only as refuges for animals and plants but undisturbed laboratories for discerning the principles embedded in complex ecological interactions.[23]

That habitat preservation by public agencies today seems so intuitive as a strategy for species protection is partly a measure of the success of scientists in making this argument at the start of the twentieth century, even though at first it faced a chilly reception from government. In 1916, for example, Grinnell and fellow UC-Berkeley researcher, Tracy Storer, suggested that the nation's existing national parks—then just reorganized under the new National Park Service—were valuable for scientific research in addition to their known recreation value. As "the settlement of the country progresses and the original aspect of nature is altered," Grinnell and Storer wrote, national parks would "probably be the only areas remaining for scientific study."[24] T. S. Palmer—the first to collect pupfish from Devils Hole in 1891 as part of the Death Valley Expedition—made a similar case a year later, telling a conference on the future of the national parks that existing monuments, such as Grand Canyon, could be "maintained as a sanctuary for wild life [and] should become practically a natural outdoor laboratory or observatory."[25]

Others took Grinnell and Storer's idea for habitat protection further. Not only should existing parks, monuments, and forests be run for the benefit of animals and plants and their study, but additional areas should be acquired for this purpose. "Large tracts of land, representing every type of physiography and of plant association, ought to be set aside as permanent preserves," wrote F. B. Sumner, the Scripps Institution of Oceanography researcher.[26] With noted ecologist Victor Shelford and others, Sumner served on the Ecological Society of America's Committee on the Preservation of Natural Conditions, which worked during the 1920s to establish a list of sites that should be protected to preserve plants and animals. "It does not seem to me utopian," Sumner concluded, to have the government set aside tracts of land for "the permanent preservation of the native fauna and flora by reason of their value to science, and to the higher interests of generations to come."[27] If not necessarily a utopian idea—President Teddy Roosevelt created the first U.S. wildlife refuge in 1903 with the reservation of a three-acre island in Florida—expanding the concept would not be easy.

Sumner, in addition to his role in mainstreaming the idea of habitat preservation for wildlife in the United States, generally, also sparked the first conversation I have uncovered about protecting habitat around Ash Meadows and Devils Hole, specifically. Sumner and a colleague, as well as their wives, traveled to Ash Meadows in 1939 armed with driving directions from Robert Miller. They undertook the first ecological research on pupfish in Ash Meadows, conducting in situ experiments on the metabolic rates of fishes living in springs with different temperatures using two populations of Ash Meadows pupfish.[28]

After another desert trip two years later to Lincoln County, Nevada, a pattern struck Sumner. He had noticed habitat alterations at a number of springs in both places, likely the result of agricultural development, which then led him to worry about the future of fishes living there. Each example of "disturbance" raised a question. He wrote, "Can anything be done about it?"[29] Answering his own question by channeling his profession's belief that habitat could be protected by federal agencies, he suggested that Ash Meadows' springs could be made into a national monument.

• Nature's Playgrounds

F. B. Sumner never returned to Ash Meadows after his 1939 trip. And, as far as I can gather, he never again mentioned in writing the possibility of making it a national monument—he died in 1945. The idea that the springs and their unique inhabitants were worthy of protected status, however, lived on through the only person he shared his musing with: Carl Hubbs, the energetic ichthyologist who supervised Robert Miller's graduate work on pupfish taxonomy.

Two years after Sumner mentioned it, Carl Hubbs suggested to the National Park Service that the agency protect Ash Meadows.[30] The area, he informed Carl Russell, Park Service chief naturalist, lay "near but outside" Death Valley National Monument and contained "very interesting springs, some of which are the habitat of certain fishes of extremely isolated range and very limited numbers." Hubbs highlighted Devils Hole, and its population of "at most a few hundred" fish, to argue that the entire area was important, not to suggest that only Devils Hole should be protected. "We have seen some of the isolated fishes of the desert pass out of existence within recent years," Hubbs continued, "and I believe the Ash Meadows group of springs would be a very logical one for Park protection."[31] He asked Russell the best way to submit a more formal proposal to the agency.

Hubbs was already well known around the Park Service when he wrote this letter. During the 1940s and 1950s, he collected fishes across the park system and authored papers on the fishes of different national parks—

including Yosemite and Isle Royale—and was frequently in touch with senior officials at the agency.[32] He also sometimes opposed National Park Service policy during this period, criticizing the stocking of nonnative sport fishes in Yosemite, but the agency continued to seek out his opinion on a range of natural resource issues.[33] Just a few weeks after Hubbs wrote to Russell about Ash Meadows, NPS Director Newton Drury solicited his thoughts on the management of Yellowstone's bison herd, with Hubbs suggesting the agency consider "the natural control of excess bison by the large predators."[34] (It would take some time for the Park Service to catch up to this view; it reintroduced wolves to Yellowstone only in 1995.) Regardless of Hubbs's criticisms, Drury told him in early 1950, "You can be always helpful if you will continue to keep a watchful eye on our biological work and tell us when you think we are wrong, as well as informing possible critics when we seem to be right," an invitation Hubbs took to heart in the Devils Hole case.[35]

Partly as a result of Hubbs's prominence among biologists and his relationship with NPS officials, several senior officials read his letter after Russell circulated it for comment within the Park Service and the Fish and Wildlife Service. His idea was not met with universal praise. Clifford Presnall, the chief of the section of wildlife research on public lands at Fish and Wildlife responded, "Personally, I'm a little cool toward such scientific minutae [*sic*] in a national system."[36] Conrad Wirth, NPS's future director, echoed Presnall, writing, "This seems to be a purely scientific matter."[37] But Victor Cahalane, a biologist, had a very different take: "Although the animals in question may be small in size and therefore regarded by the layman as insignificant, they may have great biological importance and therefore be of sufficient national interest" to protect as part of a national monument.[38] Carl Russell, the NPS naturalist, suggested that Hubbs make a "formal recommendation" to the agency's director, but spared him the less encouraging views circulating within the temporary wartime offices of the Interior Department in Chicago.[39]

The differing views from Wirth on the one hand, and Cahalane on the other, are not surprising. While they had similar backgrounds—both attended the Massachusetts Agricultural College (now University of Massachusetts, Amherst)—Wirth had pursued landscape architecture while Cahalane eventually found his way to the wildlife division of the Park Service. Moreover, their responses to Hubbs's informal proposal mirrored a central tension at the heart of the National Park Service's mission from its inception in 1916. Congress's enabling legislation stated that NPS would "conserve the scenery and the natural and historic objects and the wild life" in the parks and monuments it managed, but also that the agency should "provide for the enjoyment of the same" by visitors and "future generations."[40]

That a landscape architect wanted to improve the parks for visitors and a biologist for wildlife is not surprising. But these two views did not carry equal weight in the early Park Service. Over NPS's first decade and a half, the agency interpreted its mission primarily by developing the parks for tourists—leaning heavily on the "enjoyment" aspect of the mandate, not the conservation of "wild life" part. In a 1918 memo nominally written by Secretary of the Interior Franklin Lane, but actually authored by Park Service Assistant Director Horace Albright, the agency laid out its vision for administering the parks, calling them a "national playground system"—one with high standards. New areas should enfold "scenery of supreme and distinctive quality or some national feature so extraordinary or unique as to be of national interest and importance," and the system "should not be lowered in standard, dignity, and prestige by the inclusion of areas which express in less than the highest terms the particular class or kind of exhibit which they represent."[41]

Death Valley National Monument was added to the park system under such assumptions in February 1933, one of the last acts taken by President Herbert Hoover before the end of his term. While deserts were not considered areas of great scenic interest in the nineteenth century, by the early twentieth century that attitude was changing such that Hoover used his waning days in office to establish not only Death Valley, but two other desert monuments: Saguaro (near Tucson, Arizona) and White Sands (in southern New Mexico).[42] "The 2,500 square miles included in the monument embrace Death Valley itself and parts of the rough-hewn mountains that rise abruptly on all sides to guard its colorful desolation," a 1934 Park Service visitor brochure read.[43] Deserts, it now seemed, could also meet the Lane letter's call for "scenery of supreme and distinctive quality."

Absent in the Lane letter, as well as in the early administration of Death Valley, was much of anything about the criteria for managing the plants and animals living in Park Service areas, or as a rationale for future acquisitions. For the first decade and a half of its existence the Park Service had no resource management or ecological planning nor did the parks themselves provide a representative portrait of ecosystems in the country. Death Valley National Monument had pupfish inside its boundaries upon establishment in 1933—at Salt Creek, just north of monument headquarters, for example—but such inclusion was coincidental to the reasons it was created.

Thus, when Scripps ecologist F. B. Sumner laid out his vision for protecting habitat using public lands in 1920, he thought that while the national parks had a made a start, they often only captured "areas of exceptional scenic grandeur" that were primarily "playgrounds." What was needed, he thought, were "other tracts, in which the fauna and flora are reserved primarily for the studies of the botanist and zoologist—ones in which the native life will be more adequately safeguarded than at present."[44]

An alternative course for the Park Service, more in line with Sumner's approach, emerged from the agency's short-lived wildlife division, established in 1933. Headed by a young, energetic ecologist—and protégé of UC-Berkeley's Joseph Grinnell—named George Wright, the division attempted to establish ecological principles in the management of national parks and monuments. The division's staff proposed, among other things, evaluating new park projects—roads, buildings—with a view to their effect on the area's animals and plants, ending the practice of predator elimination, and even setting park boundaries based on the ranges of wildlife. Beyond such specific recommendations, however, the wildlife division's work suggested something fundamental: that in order to be true stewards for "future generations," the Park Service would have to invest in learning about the plants and animals living in the places it managed, something it had done little of during the 1910s and 1920s.[45]

Carl Hubbs did not have great timing. He first brought Ash Meadows and the pupfish to the attention of NPS after the power of the wildlife division to affect park policy had already waned. While NPS Director Arno Cammerer had declared the recommendations from the wildlife division's first report to be agency policy (a 1933 document called, in Park Service circles, Fauna No. 1), it was not permanently incorporated into the agency's culture. During the 1930s, the wildlife division did employ as many as twenty-seven biologists to guide park management, but most of these were hired with Civilian Conservation Corps funding, which dried up as the U.S. moved toward a war footing in 1939 and 1940. At the same time, the agency employed many more landscape architects, for whom aesthetics and visitor experience were paramount. With such an imbalance in numbers,

NPS biologists increasingly found themselves outnumbered and on the defensive.[46]

By the time that Hubbs wrote to Carl Russell and the Park Service in 1943, the Wildlife Division had even been moved out of the agency all together—Cahalane and his colleagues had been transferred to the Fish and Wildlife Service. Starting in 1944, the biologists were moved back into the Park Service but were still routinely ignored.[47] In a letter to Carl Hubbs in 1951, NPS biologist Lowell Sumner (no relation to F. B. Sumner) wrote, "For five years I have been asking for a fisheries biologist and have made progress to the point where occasionally someone else in the office also mentions the desirability of such a position. This is at least a step forward."[48]

Hubbs did not push the agency on Devils Hole and Ash Meadows again after his 1943 letter to Russell, so the issue was moot—for the time being. Later, Park Service intransigence was a major stumbling block for the protection of Devils Hole and Ash Meadows. In 1949, Hubbs acknowledged that he had "still failed to get around to specific recommendations for the preservation of some of the isolated and therefore threatened species of fish in the Great Basin region. I hope to be able to do this before very long."[49] And he had not thought to contact the Fish and Wildlife Service at all.

• Nice Weather for Ducks

In the middle decades of the twentieth century, the Fish and Wildlife Service undertook a veritable spree of habitat acquisition explicitly for the nation's wildlife. An agency that began as C. Hart Merriam's tiny Division of Economic Ornithology in the 1880s, and along the way collected the first pupfish from Devils Hole, had, by the 1940s, grown into a sprawling Interior Department agency that operated 271 separate wildlife refuges across nearly 10 million acres in the contiguous United States.[50] In 1947, it even produced a report recommending the establishment of a refuge in Ash Meadows. As "the forces of modern civilization... reduce[d] the land and water areas that are suitable for wildlife," the agency's 1946 annual report read, "a variety of land and water areas must be set apart and maintained for the primary purpose of providing for the needs of wildlife."[51]

This should have been music to Carl Hubbs's ears. Despite this expansive program, however, Hubbs never proposed using the Fish and Wildlife Service as a vehicle for protecting aquatic species and habitats in Ash Meadows or anywhere else. In his correspondence—and even in a resolution by the

American Society of Ichthyologists and Herpetologists passed at the same meeting where Hubbs became the organization's president—he repeatedly focused on the Park Service as the best chance for protecting fishes and their habitats.[52] But with NPS seemingly so uninterested, why did Hubbs overlook the possibility that the Fish and Wildlife Service could protect Ash Meadows?

It is tempting to think about this episode as a tragic missed opportunity that could have avoided much of the subsequent controversy over Devils Hole and Ash Meadows, short-circuiting decades of strife that threatened the Devils Hole pupfish with extinction. I certainly thought about this episode that way when I first uncovered the Fish and Wildlife Service records: the federal government could have bought the private land in Ash Meadows for a pittance, transferred the adjoining public land to FWS management, and "saved the pupfish" way back in the 1940s!

A closer look made me rethink this impression. I now know that Hubbs was smart—and the pupfish lucky—to avoid FWS management in the 1940s. While today the agency is the main administrator of the U.S. Endangered Species Act and operates fifty-seven wildlife refuges explicitly for the benefit of endangered species (now including Ash Meadows), in the 1940s FWS was a different animal.[53] When the agency looked out at the landscape, in Ash Meadows and around the country, all it could see were the ducks.

Despite the name—wildlife refuges—the sites that FWS managed were focused on specific animals, especially migratory birds that travel hundreds or thousands of miles between northern summer breeding grounds and southern wintering habitats. Carl Hubbs did not imagine the pupfish in a wildlife refuge because the Fish and Wildlife Service did not concern itself with endemic nongame fish like the pupfish. And while later the pupfish's survival would owe much to the Fish and Wildlife Service, in the 1940s the Devils Hole pupfish and a host of Ash Meadows endemics likely survived because the agency did *not* act on its plan.

The nation's first wildlife refuges date to Theodore Roosevelt's administration, but the creation of a real *system* of refuges began in the 1930s. Migratory birds had been declining as the result of both overhunting and habitat destruction during the late nineteenth and early twentieth centuries—the very forces that kept extinction in the minds of Sumner, Hubbs, and other scientists. Hunting restrictions had helped slow the decline, but agricultural expansion into wetlands continued to affect bird populations. In California's

Central Valley, an area critical for birds migrating along the Pacific coast, some 90 percent of the region's wetlands had been destroyed by the time the Fish and Wildlife Service (then named the U.S. Biological Survey) began acquiring refuges in the 1930s.[54] Under the leadership of Jay "Ding" Darling, the Biological Survey successfully argued that it should acquire refuges along these migratory routes, which they called "flyways," to provide sanctuaries for birds amid habitats that had been rendered unusable by agricultural expansion.

The creation of this national wildlife refuge system worked; gradually migratory bird populations traveling along the flyways bounced back. But the birds did not stop over or winter at pristine, federally owned wetland habitat. Instead, the Fish and Wildlife Service created and intensively manipulated the landscape, producing something more like "duck farms" than the natural areas that ecologists such as F.B. Sumner wanted protected.[55] When Ash Meadows got its closest inspection from the Fish and Wildlife Service, in 1947, Doren Woodward, a supervisor in the FWS's division of lands, knew what creating a refuge would require. After suggesting acquiring 7,980 acres for this purpose, Woodward remarked on Ash Meadows' central advantage: the meadows were "wet and could readily be broken into units for water control."[56] Managers routinely created dikes and ponds at refuges to manipulate water levels, limit the spread of avian diseases, and even grow crops to feed waterfowl. Ash Meadows might have been a good spot for a refuge, but Fish and Wildlife Service managers would have to radically alter its network of springs to improve the area's bird habitat.

Woodward's report was enthusiastically seconded by the manager at the FWS's Desert Game Range north of Las Vegas who noted its "exceptional qualifications as a refuge area greatly needed to protect the dwindling flight line of ducks in this locality." The manager, Frank Groves, would later become the director of the Nevada Department of Fish and Game. He too envisioned manipulating the springs and their outflows, creating ponds to make the area more attractive to birds. The water, he wrote, "could very readily be impounded without any great expense." Furthermore, Groves thought that at least one spring in Ash Meadows could also be used as a warm-water hatchery for sport fishes.[57] Neither Woodward nor Groves mentioned the pupfish (or any of Ash Meadows' other endemic species), yet impounding water behind dams and introducing exotic fishes, if not immediately detrimental to the pupfish in Devils Hole, certainly boded ill for Ash Meadows' other aquatic species. As geographer Robert M. Wilson has noted in his history of wildlife refuges in California and Oregon, "What was good for waterfowl was not necessarily good for other species."[58]

Still, given such support from Fish and Wildlife Service employees, why did Ash Meadows not become part of this expanding network of migratory bird refuges? Two possible answers emerge from the record. First, the agency resorted to the classic explanation of all government agencies when they do not want to do something: it said it did not have the funds. The "present financial situation," the Washington office said, prevented action for "some years to come" and told Woodward and Groves "no further examination work should be done on the project."[59] In this explanation, tight budgets saved Ash Meadows and the Devils Hole pupfish from the Fish and Wildlife Service.

A second, more circumstantial explanation of the Fish and Wildlife Service decision emerges from investigating who in the agency spiked the proposal: J. Clark Salyer II, chief of the refuge division and a former student of Carl Hubbs at the University of Michigan.[60] Salyer had even visited Ash Meadows when the Biological Survey first evaluated it in the 1930s.[61] In 1947, Salyer rejected Woodward's plan by scribbling a brief note: "I don't see any chance of considering this area for some years to come. I have seen it previously."[62] Salyer kept his reasons to himself; it was Salyer's subordinate who introduced the financial argument in communicating with the agency's Nevada employees. I like to think that Salyer, having seen Ash Meadows, and knowing full well the kind of transformation that creating a bird refuge required, perhaps resisted turning the area over to his own agency out of deference to the endemic fishes loved by his former teacher. This last bit is pure speculation, but what I do know is this: three years later, when Carl Hubbs finally began pushing the Park Service to acquire Ash Meadows and Devils Hole, J. Clark Salyer was an eager supporter.[63]

• Protect from Whom?

The National Park Service was lukewarm on protecting the Ash Meadows spring habitats. And the Fish and Wildlife Service missed the desert fishes entirely when it evaluated this landscape; it could only envision improving the area for birds. University scientists, however, knew what they had right in front of them: a stunning assemblage of endemic fishes in danger of being wiped off the earth and thus worth protecting.[64] They also almost killed them.

At Devils Hole, the study of the pupfish—the very undertaking that defined the species and spurred conservation efforts that enabled its survival—may also have been the greatest threat to their persistence during the 1940s and early 1950s. Remember that when Robert R. Miller visited Devils Hole in June 1937, he told Hubbs that the pupfish was in trouble. He estimated the population at "not more than fifty or sixty fish." In another letter, he said it was forty—and believed it would not be long until it was extinct. Despite this belief, however, Miller returned to Berkeley with twenty-eight Devils Hole pupfish preserved in alcohol.[65] Based on his own account, Miller removed 50—perhaps even 70—percent of all the Devils Hole pupfish in the world from their habitat. While vigilant for the effects of habitat loss or hunting on species, scientists could also put their research subjects at risk.

Miller did not hold the record for pupfish collecting. In his 1930 paper that identified the Devils Hole pupfish as a unique species and gave it the name *Cyprinodon diabolis*, Joseph Wales referenced having taken 60 specimens from the habitat, while estimating the total population at about 200.[66] Before he analyzed the fish from Devils Hole, the population was not understood to be a unique species, just a population of a fish common to the U.S. West. After publishing his paper, however, the scope of the collection seemed more significant. Many of Wales's preserved fish found their way to the Smithsonian's collections via his colleague, George S. Myers, who had accompanied him to Devils Hole in 1930. In the Smithsonian catalogue, Myers suggested that the 42 fish the museum held represented "about one third the existing population of this species at the time the collection was made!!"[67] Decades later, when a pupfish researcher asked Wales about how many fish he removed from Devils Hole, he reportedly replied, "You don't want to know."[68]

As discussed in chapter 1, collecting fish determined how species themselves were defined. When taxonomists like Charles Gilbert, Joseph Wales, and Robert Miller went about the task of grouping the pupfishes of Ash Meadows and Death Valley, they were essentially engaged in a statistical exercise. The greater the number of fishes from each desert habitat available, the easier it was to tell whether different measurements of size and shape were features of aberrant individuals or defining characteristics common to a subspecies, a species, or a genus. Hubbs told Miller before his 1937 trip to Devils Hole that collecting a large number of fish was "very desirable," because a "rather large series... allow[s] statistical study of raciation."[69] Even after defining a species, though, scientists typically continued collecting "voucher specimens"—additional specimens that could be used to answer new questions or simply provide documentation of a species' existence in a locality at a specific moment.[70] Essentially, the science of species and the practices of natural history museums depended on dead bodies.

For species with very large populations, it was unlikely that scientific collecting could negatively affect the whole. Presently, animal rights activists might find collections morally objectionable—at the very least, natural history museums themselves can be somewhat macabre—but at a population level, collections are mostly benign. Yet for stressed or isolated populations of animals, "overcollecting" could be catastrophic. Collecting too many individuals from a population might lead to reduced genetic diversity among those remaining, producing the long-term problem of inbreeding depression where offspring suffer from maladies that make them less successful in their environment and less likely to reproduce. Overcollecting may also make a species more vulnerable to other kinds of threats. The California condor, for example, was subject to extensive collecting during the nineteenth and early twentieth centuries. In combination with subsequent problems—lead and DDT poisoning, for example—this collecting played a role in the decline of the species in segments of its historical range.[71]

With small populations, there is also the risk that scientific collectors may actually capture the final individuals in a species, which appears to have occurred in the case of the great auk, a species of large flightless bird native to rocky islands ringing the north Atlantic. Pursued for its meat, oil, eggs, and feathers beginning in the seventeenth century, in 1844, collectors looking to sell specimens of the then rare (and thus valuable) bird to natural history museums collected the last two known individuals of the species.[72] Many other lesser-known species have suffered, or nearly suffered, the same fate from scientific collectors.[73]

In total, between 1930 and 1941, at least 203 specimens were taken from Devils Hole and preserved at natural history museums around the country.[74] Luckily for the pupfish, Miller, Wales, and Myers all likely underestimated the Devils Hole population size. Estimates made near the surface necessarily missed the pupfish that descend deeper in the water column to hide when disturbed, thus even if Miller saw 50 or 60 fish in 1937 the population was probably much larger. As discussed later in this book, when scientists began regular scuba population counting in the 1970s, they have observed that Devils Hole pupfish populations vary seasonally, since during the winter months no sunlight reaches the pool in Devils Hole and thus limits the amount of pupfish food being created. The population usually reached a maximum population in the fall and a minimum in the spring. During the 1980s, for example, pupfish tallies during the fall ranged from 396 to 548; during the spring it ranged from 143 to 237 individuals.[75] So although Wales and Miller's collecting was substantial during the 1930s and early 1940s, they likely collected a smaller percentage of the population than they believed.

The closest scientists came to destroying the pupfish, ironically, was within days of the first rigorous effort to measure its population. Perhaps unsatisfied with his previous estimates of the population size of the Devils Hole pupfish, on April 1, 1950, Robert Miller returned to Devils Hole and conducted the first complete census of the pupfish population by attempting to remove all the fish from the pool with a seine and temporarily holding them in a tank. Miller was able to assemble about 88 fish this way, and then still could see 22–24 adults remaining in the pool. He made an additional estimate of 20 young still in the pool and then added 10 percent for those fish he did not see, bringing the total to 140, "or about 150" fish.[76] He returned the fish to the pool. Miller's method still had a certain back-of-the-envelope quality—he actually recorded the tally on the inside cover of his field notebook—but it was the first relatively rigorous estimate of the population, and his number sits on the low end of population estimates recorded in the pool by the scuba-diving method during the 1980s.

Ten days after Miller's visit, on April 10, 1950, in an apparent coincidence, a scientist named Ira La Rivers arrived at Devils Hole. La Rivers was a University of Nevada professor and a contributor to the collections at the university's small natural history museum. Trained at the University of California as an entomologist, once in Nevada he became fascinated with the state's fishes, eventually publishing a lengthy tome titled *Fishes and Fisheries of Nevada*.[77] (Miller called the book "a most ambitious undertaking—perhaps too much so for practical purposes."[78]) La Rivers removed and preserved 74 pupfish from Devils Hole.[79] In the span of 10 days, one scientist made the most reliable estimate of Devils Hole pupfish population yet, and then another scientist removed and preserved a full half of the known population.

La Rivers visited Devils Hole more than once. In fact, between 1946 and 1951, he and his assistants visited Devils Hole five times, removing hundreds of pupfish from the habitat. From these visits—all of which occurred during winter or spring, when pupfish populations are typically lowest—La Rivers accumulated the largest single collection of Devils Hole pupfish in the world: more than 200 preserved specimens, which remain at the University of Nevada to this day. But determining La Rivers's impact on Devils Hole by looking at natural history collections does not tell the whole story. It only offers documentation of the fish that made it onto storage shelves. There is more. On the first of his five known visits to Devils Hole, in the winter of 1946–1947, La Rivers removed 75 live Devils Hole pupfish and placed them in a spring in Ash Meadows three miles to the west.[80] On that same trip, La Rivers also preserved 33 specimens for the natural history museum at the University of Nevada, meaning he actually removed 108 fish from Devils Hole. It is possible that La Rivers removed two-thirds of all pupfish from Devils Hole at once.

It took me a long time to piece together this story of La Rivers's collecting and transplanting, since the evidence was scattered across natural history museum databases, old journals, and archives. The extended period of research meant there was not a single moment where I gasped in realization that his work may have seriously harmed the species, potentially influencing the pupfish's gene pool or altering its population structure through the removal of so many adults. I have, however, been troubled by the loose ends that my research left me holding, because what I uncovered suggests that there may be more we do not know about La Rivers's actions at Devils Hole. For instance, there may have been additional pupfish transplants. The only accounts of his 1946–1947 pupfish transplant are in two obscure journal articles published years after the event.[81] La Rivers's personal papers—unlike those of Robert Miller and Carl Hubbs, which I draw on throughout this book—appear not to have been preserved after his death in the 1970s. The California Academy of Sciences in San Francisco does have one small set of materials related to his research on invertebrates but not much else. I decided to go and sift through them anyway with the hope that there might be a stray piece of paper related to the pupfish among them. The most significant thing I discovered, though, was a note from a former museum employee who wrote of La Rivers's papers that "one son may have discarded and thrown away other files."[82] This, I reflected as I looked over five boxes of material related to his work on a group of tiny insects called naucorids, was a real loss. The organization of the records in front of me revealed the care he took in his research. I can only imagine what a full accounting of his pupfish collection trips would disclose.

Despite this gap, there are two tantalizing hints that the 75 pupfish he reported transplanting may be just the tip of the iceberg. In a 1958 letter, La Rivers told Robert Miller that "as far as we are aware none of our early transfers of *Cyprinodon diabolis* were successful."[83] Note the plural: "transfers." DNA research on pupfishes in the Death Valley region may also provide documentation of these efforts. In a 2016 study, a group of geneticists found evidence for Devils Hole pupfish genes in the DNA of two other pupfish species. Their method also suggested that this "introgression" of Devils Hole genes occurred in the recent past.[84] Since pupfish species in the Death Valley region readily hybridize when introduced into the same habitats, perhaps the La Rivers transplants explain this evidence.

Another question that emerges is why La Rivers decided to transplant and collect so many fish. In his publication detailing the 1946–1947 transplant, he offered a counterintuitive answer: conservation. La Rivers wanted to "provide some sanctuary for the unique species," feeling that unspecified "tampering" with Devils Hole by "local interests would result in extermination of the species."[85] Yet not only did La Rivers's own actions endanger the pupfish by removing substantial portions of the population, moving them to habitats with other pupfish species also meant that the species would not be conserved in the way he imagined. While he embodied his profession's concern over the future of species in arid environments, his conservation vigilantism was ham-fisted and dangerous.

La Rivers's behavior alarmed Carl Hubbs. In January 1952, Hubbs learned of the 75 transplanted fish—though not of the more numerous preserved specimens—and chastised La Rivers: "I do hope that in any future such attempts to save the species you do not seriously decimate the home stock!" While collecting was essential to science, Hubbs had already worried that "excessive collecting" was a threat facing the pupfish and La Rivers's actions seemed to substantiate his concern.[86] Hubbs told La Rivers that once the Park Service began managing the site (then imminent), such individual approaches to conservation would be unnecessary and that only a "qualified person" would be given access.[87] Unsaid was whether Hubbs would consider La Rivers qualified.

After Death Valley National Monument took over at Devils Hole, Hubbs recommended that no more collecting occur. "Quite adequate collections are already available... so that there should be no real need for collecting additional specimens," he told monument staff. There was also, he said, no point "in trying to introduce them into other waters, as was done some time ago."[88]

• Love Letters

At the very end of William Golding's classic novel, *Lord of the Flies*, a naval officer suddenly appears out of nowhere on the deserted island where a group of schoolboys has been marooned, and all the internally generated action in the story—the conch, the pig head on the stick, an impending murder—comes to a screeching halt. As every tenth grader who reads this in high school English class is told, the unexpected redirection or conclusion of a story by an outside force is a plot device called the *deus ex machina*, god from the machine.

While the pupfish story does not (hopefully) share much with Golding's tale—among other things, this book is a work of history, not fiction—the path of the Devils Hole pupfish to the national park system began with a kind of *deus ex machina*. To varying degrees, the Park Service, Fish and Wildlife Service, and a small group of scientists had all looked at and studied the Devils Hole and Ash Meadows landscape during the 1940s. The closest anyone had come to protecting the Devils Hole pupfish was Ira La Rivers—and he almost killed them. The pupfish survived these aborted plans and ill-conceived preservation schemes. Then, in June 1950, an official report from yet another federal agency—the U.S. Bureau of Land Management (BLM)—arrived at Park Service headquarters. BLM asked whether NPS would like to reserve Devils Hole for a possible national monument.

How exactly BLM learned of Devils Hole and proposed that the Park Service take control is worth considering. But like a good deus ex machina, most of this will be left unstated for the moment. BLM is an Interior Department agency with responsibility for managing the vast public domain not contained in military bases, national parks, monuments, forests, or wildlife refuges. Referred to occasionally as the "nation's largest landlord," at its establishment in 1946 BLM was especially focused on overseeing mineral rights and grazing leases. Species protection was not in its purview until much later in its history.[89] BLM's reasons for proposing the site to the Park

Service for a monument is discussed in chapter 3, because it has almost nothing to do with land or fish, and everything to do with water. There is no evidence that NPS ever knew why BLM proposed it. For now it is enough to know that a land appraiser for the agency, Jean M. F. Dubois produced a report, given the identifier SF94949, which noted that the scientists he contacted—including Ira La Rivers—supported the idea. Dubois then suggested that 40 acres "be reserved as a national monument... and withdrawn from all forms of entry and appropriation including mining and mineral leasing in order to protect the scientific values of the Devils Hole, located thereon."[90] BLM transmitted Dubois's report to NPS and asked if it would like to reserve the piece of property. The agency's investigation also came with a caveat: it only examined the area around Devils Hole—not any of Ash Meadows' other springs.

During the summer and fall of 1950, the Park Service evaluated Devils Hole, replaying many of the sentiments from years earlier when Hubbs wrote to the agency about Ash Meadows, but this time only focusing on the one habitat identified in the BLM report. It also replayed the decades-long tension over the purposes of the park system. NPS director Newton Drury told BLM that Devils Hole "lack[ed] national significance" and that NPS did not believe it could justify a national monument on the grounds of only "animal life."[91] Still, Drury and agency leaders wanted to be thorough before officially declining BLM's offer, so they asked Death Valley officials to submit reports on the proposal. "Aside from the claimed fact that *Cyprinodon diabolis* are a distinctive species," Death Valley Superintendent Theodore Goodwin reported back after visiting the site, "it hardly seems that there is sufficient value to consider it for a National Monument."[92] The monument's naturalist concurred.[93] Ultimately, by October 1950, Drury summarized the agency's views in a letter to BLM: Devils Hole was "of very real scientific interest, [but] it is felt that it does not possess qualifications of national significance sufficient to warrant its inclusion in the National Park System."[94]

From his post at the Scripps Institution in La Jolla, Carl Hubbs seems never to have been aware that the Fish and Wildlife Service toyed with transforming Ash Meadows into a migratory bird refuge. The same would likely have been true for the Park Service's October 1950 decision to pass on Devils

Hole, but for the fact that Hubbs had an inside man at Death Valley: his son, Earl, worked in the monument store.

When Hubbs and his wife visited Earl in Death Valley in November, he learned of NPS's decision and paid a call on the monument's naturalist, Floyd Keller. Keller chatted with Hubbs and apparently even loaned him the Park Service's record of letters related to BLM's proposal for Devils Hole. Hubbs did not like what he found, and his response constituted both a rejection of the priorities of the Park Service on display in its characterization of this desert habitat and a major turning point in the fortune of the Devils Hole pupfish.[95]

Carl Hubbs's rejoinder to the Park Service's decision came in the form of a letter to its director, Newton Drury, mailed just days after returning to La Jolla. Playing off of the 1918 description of the national parks as a "national playground system," and echoing F. B. Sumner's critique of the Park Service from the 1920s, Hubbs wrote, "Perhaps I have been naïve in assuming that preservation of nature was among the basic reasons for and functions of the National Park Service," adding, "I would hate to think of your department as only a National Playground Service." Instead, Hubbs told Drury, "It *is* a national concern to preserve a habitat and a species as unique as are [*sic*] Devils Hole and its endemic fish...even though the preservation be primarily for scientists." Hubbs then argued that the range of threats that had led to other extinctions in the West could be visited on Devils Hole—water development, invasive species introductions, or by "excessive collecting." He concluded that Devils Hole *and* other springs in Ash Meadows should be managed by the Park Service.[96]

It appears that Drury did not reply to Hubbs's letter. But Drury was not the only intended audience. Copied on the message were eleven other individuals and offices from across the Park Service, other federal agencies, and academia. Hubbs was well connected in all three areas, and, as he repeated to Keller, the Death Valley naturalist who had shared the monument's correspondence file, "I will try and carry on, spreading interest in the matter, until something is accomplished."[97]

It did not take long for interest in the matter to come back to Hubbs. He quickly assembled considerable support for protecting Ash Meadows under NPS management. J. Clark Salyer—the official at FWS who had declined to pursue remaking Ash Meadows into a national wildlife refuge for migratory waterfowl—replied to Hubbs, writing he was "very sympathetic with the case you present with respect to *Cyprinodon diabolis*."[98] Salyer also informed some other heavy hitters in the conservation world: Ira Gabrielson (the former head of the Fish and Wildlife Service), Howard Zahniser (a founder of the Wilderness Society), and officials at the National Parks Association, an advocacy group.[99] The Parks Association, in particular, began to investigate Hubbs's proposal, including the possibility of protecting other springs in Ash Meadows containing endemic fish.

Dissatisfaction also bubbled up within the Park Service over its Devils Hole decision. As a result of Hubbs's letter, the agency's regional office sent Lowell Sumner (no relation to the Scripps ecologist) to Devils Hole to reexamine the area, since he "was to be in the neighborhood in December" anyway.[100] Lowell Sumner was one of the few biologists employed by NPS, an acquaintance of Hubbs, and a veteran of NPS's brief foray into ecological management as part of its 1930s wildlife division. That is, he was exactly the kind of employee who might question the judgment of Drury and the Death Valley officials.[101]

Sumner crafted a new evaluation, one very different from those presented to NPS leadership by Death Valley employees the previous summer. His eight-page report, complete with photographs, began by summarizing the existing arguments made by his employer for not acquiring Devils Hole.[102] "Service thinking," Sumner wrote, suggested three things: first, that some other agency "would be willing to protect it;" second, that the Park Service did not have the authority to acquire it; and third, "we are not yet convinced that it meets our standards." He then offered a rebuttal. On the first two counts, Sumner said, "to be realistic," there was no other agency to provide protection and "it would seem that incorporating Devils Hole into Death Valley National Monument as a detached portion ought not to present any insuperable difficulties if it meets Service standards." As for whether it met those standards, he said yes, both on scientific grounds ("so much so that we need not argue the point") and scenic beauty (though "this depends more on each individual's point of view," he conceded). The passage of time has made Sumner's analysis seem intuitive. In fact, many people might assume this is the attitude that NPS has always brought to biological resource issues, instead of an argument that had to be made. Sumner also believed that acquiring Devils Hole would "not be opposed by anyone." (As it turned out, not for at least twenty years, anyway.)

The efforts of Hubbs and Sumner went beyond their critiques of NPS or their specific proposals. Each was a kind of love letter to the pupfish, dwelling on the marvel of its existence, as they argued for a different kind of Park Service, one that would see pupfish as central to its preservation mission. The pupfish were "absolutely unique," Hubbs told Drury. Standing at the foot of the pool, "one could see at one time more than 90 percent of all individuals comprising the entire species." "I know of no match for this situation," he wrote.[103]

For his part, Sumner opined, "If you love the desert, there is considerable charm in this quiet, mysterious home with its clear, warm water welling up from deep underground, and the swarming little fish that long ago found refuge here, fifty feet below the surface of the earth." Acquiring the area, he believed, would also "justify the faith of scientists...that, with respect to areas of national importance, we are indeed a conservation Service." Looking into Devils Hole provided a sense of wonder at the journey that the pupfish had traveled to the twentieth century, and by protecting it the National Park Service could also save itself.[104]

Hubbs and Sumner won. In March 1951—just four months after Hubbs penned his letter to NPS—the Park Service made an about-face and announced that it had reconsidered its position, recommending that Devils Hole be made a part of Death Valley National Monument.[105] National monuments are created under the Antiquities Act of 1906, which allows the president to reserve public domain land for the protection of "historic landmarks, historic and prehistoric structures, and other objects of historic or scientific interest."[106] Part of the NPS objection to Devils Hole had stemmed from a narrow reading of the act, with agency leaders believing that they could not justify a monument solely on the basis of biological life. But with criticism mounting, NPS leadership now cited Sumner's report and its observation that Devils Hole was "a remnant and remainder of the great prehistoric lakes of the Death Valley system," which, "coupled with the scientific importance" of the pupfish, justified its addition to Death Valley National Monument.[107]

In January 1952, about a year after Sumner's report began circulating within NPS, President Truman signed Proclamation 2961, adding the 40 acres around Devils Hole to Death Valley National Monument.[108] Truman's signature was apparently the easiest part of the process—he signed it without any debate, as far as I have been able to uncover, and the act went unmentioned in his daily calendar.[109] The protection of Devils Hole happened, instead, as the result of inspiration from BLM and then from pressure from outside NPS and within it. Hubbs, the outsider, and Lowell Sumner, the insider, had pushed back against the Park Service's first instinct. Reflecting on Hubbs's involvement in changing the park system, Sumner thanked him. "I think that is real democracy," he wrote.[110]

• Off the Map

Hubbs and Sumner played pivotal roles in a profound conservation victory, one that remains critical to the survival of the Devils Hole pupfish. But a wider view of the debate over Devils Hole and the eventual signing of Proclamation 2961 shows the outcome was also a partial defeat. Following the framing by BLM, the Park Service did not consider any of the other springs in Ash Meadows. Instead of encompassing a number of habitats, as Hubbs hoped and as the National Parks Association had begun to investigate, the agencies never considered it because the land was already in private ownership. The land around Devils Hole, as the BLM appraiser stated plainly, was, by contrast, "too rough and mountainous for cultivation" and so would not be missed if it became part of a national monument.[111] In part, then, the Devils Hole pupfish became protected because the land around it could not be farmed. The same was not true for the habitats of many Ash Meadows endemic species, including the poolfish, which apparently became extinct in these years. When notified that Devils Hole had been added to Death Valley, Hubbs found himself "a bit disappointed that the area could not have been made a little larger so as to contain several of the other springs."[112]

More than just disappointing, the decision to only protect the land around Devils Hole had enduring consequences, opening up a gulf between Devils Hole and the surrounding landscape. On the one hand, Devils Hole was a place of "national significance"; on the other, Ash Meadows was a place that could be transformed, if not into a bird refuge, then perhaps a large cattle ranch or a subdivision. The Devils Hole pupfish survived by being on the right side of that distinction, but in the decades since its addition to the monument, changes around Ash Meadows showed how tenuous the difference between the protected and the exploited could be.

Until the late 1960s, however, this status did not matter a great deal; the Devils Hole pupfish survived as much despite Park Service protection as from it. Death Valley forgot to mention Devils Hole in its 1952 annual report to Park Service headquarters, and it contacted mapmakers to request they not include it on their prints in order to discourage visitation.[113] Death Valley also kept Devils Hole off its own visitor maps for the same reason. The area around Devils Hole remained unfenced until 1965; NPS only installed one after two divers sneaked in and drowned there.[114]

The Park Service also failed the pupfish another way: it did not bother to learn anything about it. The agency conducted no research on the fish or the habitat, something that the NPS wildlife division proposed for all park resources back in the 1930s. Hubbs and Miller's research also moved on to other topics. Scientists, the group that had defined the pupfish as a species, made it an object of conservation, and put it at risk by removing fish from the habitat, now ignored it. By 1965, an acting assistant director at NPS laid out the gulf between the agency's charge at Devils Hole—to protect the species—and its lack of even basic knowledge. "The primary responsibility with respect to Devils Hole," he wrote, "is, of course, the perpetuation of a unique species. We do not yet know what environmental conditions are required for its survival."[115] Despite having been at the center of a serious disagreement regarding the purpose of the national park system, there was a small base of research on the fish and its strange habitat.

About the only new scientific information gained after the addition of Devils Hole to Death Valley came from regular recording of the pool's water level, initiated by the U.S. Geological Survey (USGS). In 1953, Death Valley reported that a "gauge has been installed in the Devils Hole, assumed to be by USGS employees...who called in February."[116] The sentence made clear exactly how important water-level measurements seemed to the Park Service at the time, but by the 1960s, the USGS's continuously operating water recorder had captured the most significant data at Devils Hole since Robert Miller finished counting pupfish scales in the early 1940s. The water recorder registered a threat unforeseen by NPS, Hubbs, or virtually any other individual or group involved in the battle over whether to add Devils Hole to Death Valley: groundwater pumping.

CHAPTER 3

Beneficial Use

Just about every person described so far in this book spent some time thinking about pupfish. T. S. Palmer put them in alcohol. Charles Gilbert counted their scales. Joseph Wales wondered what forces governed their different appearances. And Carl Hubbs and Park Service employees quarreled over who should protect them and how. These people did not always come to the same conclusions, but they took field notes, wrote memos, and published scientific articles all focused on these tiny desert dwellers. Robert Miller could not get them out of his head even when he slept, writing in his journal in 1938 about a "peculiar dream" where pupfish specimens he collected were "too long and slender for this genus."[1]

Most people, it probably goes without saying, do not dream about pupfish.[2] Even as pupfish became an intriguing subject for scientists and an object of conservation for the National Park Service, they remained obscure to many of the desert's human residents—as well as overlooked by Nevada's natural resource laws. In one of Carl Hubbs's first letters to Robert Miller, he cautioned his young protégé that when he stopped at a ranch or store in the desert to ask whether fish lived in any local springs, he should be sure that his "informer realizes you include 'minnows' as fishes."[3] Locals might not think of tiny pupfish as fish—a potential resource—and thus not mention it to an inquisitive student. The pupfish were out of sight, out of mind.

This chapter, in contrast to previous ones, is centered on westerners who did not dream about pupfish. It is about people in pursuit of a much more widely shared, collective dream: putting water, in the words of Nevada state law, to "beneficial use" (fig. 3.1). Instead of salt grass and ash trees, politicians and farmers saw these arid lands as future alfalfa fields tended by small farmers. In place of meadows, developers envisioned metropolises. And rather than sleepy streams, engineers sketched dams. The pupfish may have been invisible to Miller's and Hubbs's informants and many others, but the water they swam in was of considerable interest. And during the twentieth century, not just a few people noticed that the water in Devils Hole was

FIGURE 3.1. Government wildlife and land managers stand next to a well and pump installed just 900 feet from Devils Hole, 1969. Photograph by and courtesy of Phil Pister.

"unappropriated." The pupfish, then, needed to survive those dreamers—and their institutions—who missed the forest for the trees, the pupfish for the water.

Dreaming about water is not a story unique to Devils Hole or Ash Meadows. Westerners have long displayed an ingenuity for imagining oases in deserts, coming up with creative ideas for building farms and cities seemingly in inverse proportion to water's abundance. And to be sure, there is not much water to go around. In the nineteenth century, some Americans believed that "rain followed the plow" and that the climate would become more favorable as Euro-American settlement advanced west. John Wesley Powell, the second head of the U.S. Geological Survey, knew this was wishful thinking. Even if farmers succeeded in using all the available water from all the region's rivers, Powell performed some simple calculations that suggested they could only succeed in reclaiming a tiny fraction of land.[4] In the words of noted historian Walter Prescott Webb, the heart of the U.S. West is a desert, "unqualified and absolute."[5]

Despite, or more to the point, because of this reality, the scale of projects imagined and built in the west to use its scarce water are mind-boggling.[6]

In California—a state that is today home to 1,400 dams—the city of Los Angeles receives water from as far as the Mono Basin (more than 300 miles away) and through a separate system also draws water from north of the Bay Area (more than 400 miles away).[7] Moving water to southern California via this latter system requires operating 23 massive pumping plants, including one that moves an artificial river 2,000 feet uphill and over the Tehachapi Mountains. Arizona, meanwhile, pumps water from the lower Colorado River on its western border and delivers it to farms as well as Phoenix and Tucson through a 335-mile aqueduct.[8]

Among western states, Nevada holds the superlative for most arid, with a statewide average of just 10 inches of precipitation per year.[9] It is also home to some amazing feats of water imagination, serving as a test bed for federally subsidized projects to "reclaim" the desert. The Reclamation Service, established in 1902 and later renamed the U.S. Bureau of Reclamation, built a sprawling network of dams, canals, and channels east of Reno during the 1910s, irrigating some 72,000 acres in the newly formed Truckee-Carson Irrigation District.[10] Southern Nevada, meanwhile, hosts the Hoover Dam—at the time of its completion in 1935 the tallest dam in the world—and has used its legal apportionment of the water stored in Lake Mead to turn Las Vegas into a sprawling mirage of green lawns, golf courses, and backyard pools.

The Devils Hole pupfish survived the first half of the twentieth century, in part, because no pumps, aqueducts, or other water infrastructure projects materialized at Devils Hole, though there were a few attempts to utilize its water. Other species were not as fortunate. The Las Vegas dace, a small springfish confined to pools at the heart of its eponymous city, for example, became extinct by the 1950s, as groundwater pumping reduced these springs until they dried up. State laws for using water ignored the requirements of dace and pupfish equally, but one happened to live near a booming population center with demand for water, the other did not. The pupfish endured on a twist of fate, beneficiary of industrial and agricultural developments that did not occur.

As many visitors to Las Vegas well know, luck rarely lasts. One might expect that whatever risks the Devils Hole pupfish faced from water users (as well as from scientists and other federal agencies) would disappear after President Truman added Devils Hole to Death Valley National Monument in 1952. Instead, the Devils Hole pupfish became more at risk of extinction in the decades after it became part of the park system, as a large cattle ranching operation began developing water and land on a scale not seen before in Ash Meadows. The future for the pupfish became so bleak that by the summer of 1970 an NBC documentary flashed images of Devils Hole pupfish across the screen as narrator Jack Lemmon told viewers to "take a good look.... Man is about to murder them."[11]

That a ranch in Ash Meadows could drive pupfish to the edge of extinction—after already being protected by the National Park Service—was less premeditated, first-degree murder, as the star of the 1959 film *Some Like it Hot* claimed, and more like involuntary manslaughter. Yes, "man," was to blame for the population's precariousness, but the cause was not a willful system of destruction. Instead, pupfish—and other animals—were invisible to the laws governing water and land use in Nevada. The history of state water law and federal public land policy in Ash Meadows—the key developments this chapter explores—meant that the most important criterion for water and land was whether it could be put to "beneficial use." Pupfish preservation was not a beneficial use, a fact that eventually caught up to the species and its advocates in the late 1960s.

• Nothing That Could Be Done about It

I always picture John Bradford and his partner, James Coleman, standing at the lip of Devils Hole, near where the NPS observation platform is today, scratching their chins and swiveling their heads back and forth as a plan came together: 50 feet below them, the pool of Devils Hole; down the hill, the Ash Meadows patchwork of scrub, grass, and trees framed by the hazy backdrop of the Funeral Mountains. Back and forth. Back and forth.

After a minute (an hour? a week?), it must have clicked. Thanks to a peculiarity of the region's hydrogeology, the water level in Devils Hole sits well above the rest of Ash Meadows, like a water tower perched above a small town. If they dug horizontally into the side of the hill sloping away from Devils Hole, they would tunnel directly into the Devils Hole pool and then gravity would do the rest, irrigating the desert by sending the water, uselessly tied up in this cavern, rushing through pipes and ditches and onto thirsty crops.

But that image of the lightbulb clicking on over the heads of these two desert dreamers is my imagination run wild. The only details I know for sure are this: in 1914, Coleman, Bradford, and a third partner they cut in later, Charles Diehl, became the first people to attempt to get water out of Devils Hole and apply it to irrigated agriculture. Bradford was a fifty-four-year-old homesteader who had been living with his wife and four children in Ash Meadows for about four years as part of a wave of settlers granted homesteads in the area during the 1910s and 1920s.[12] Coleman and Diehl received their mail in Goldfield, Nevada, suggesting, along with scattered newspaper references, that their line of work was in the mining industry, hoping to strike it rich in the northern Nevada booms. Now they would try their luck on Devils Hole.[13]

Bradford and his partners planned to irrigate nearly 2,000 acres in Ash Meadows with the Devils Hole water, a massive project in comparison to any other existing farming operation in the area. After years of working his land, for example, Bradford had brought just 45 acres into cultivation using spring water south of Devils Hole, according to the records he filed to certify his 160-acre homestead.[14] In addition to the large size of the new undertaking, it was also expensive: the partners believed the tunnel and ditches needed to bring Devils Hole water to these lands would cost $5,000, or about $130,000 in 2020 dollars.[15]

By the summer of 1914, Bradford had made a down payment on the project, completing what he estimated was fifty dollars' worth of work by digging what he described as an "open cut and shaft" where the tunnel to Devils Hole would begin.[16] I searched for evidence of Bradford's cut in aerial photographs from the 1940s, when the first such flights recorded the Ash Meadows landscape. And then I went to have a look for myself. Enlisting my friend Brian in the search, we walked across the desert on an August morning—it was already over 90°F by 9 a.m. To the south-southwest, the distant Funeral Mountains were nearly invisible due to wildfires burning in California. We were guided by a geolocated map I had made after noting a couple of odd features in old aerial photographs that looked like they could have been Bradford's one-hundred-year-old work.

After a short walk, we stumbled across a 65-foot-long shallow ditch, oriented directly toward Devils Hole. At the end closest to Devils Hole, there was a deep opening, big enough for a person to stand in, but not much more. This, it seems, was the remains of Bradford's "open cut and shaft" (fig. 3.2). On this scorching hot morning, what was most notable about his work—aside from the fact that it is almost entirely invisible on the landscape unless guided by air photos—is how Devils Hole seemed a long 2,000 feet away.

FIGURE 3.2. Evidence of John Bradford's 1914 effort to tunnel into Devils Hole, 2018. My friend Brian (for scale) stands next to deeper hole dug at north end of the ditch. The fencing around Devils Hole is visible as a black smudge near the base of the hill descending from left side of image.

Bradford may have had the same feeling—how will I ever get all the way there?—since this was the first and last work done on the venture. Bradford's dream was larger than his ability to bring it to fruition.[17]

Completion of the tunnel in 1914 might have put an end to the pupfish in Devils Hole more than a decade before being described by the young researcher Joseph Wales, since, as the Park Service later learned, even a modest reduction in the water level could expose the shallow rock shelf on which the pupfish breed. Had Bradford's plan succeeded, in other words, the Death Valley Expedition's ten specimens sitting on the shelf at the Smithsonian's Suitland facility in Maryland would be all that was known of the Devils Hole pupfish. In fact, the fish would not have even received a unique name. Instead, Bradford's role in the Devils Hole story is quite different than he may have thought when he conceived of tapping its water supply. Years later, in 1930, Bradford directed Wales to Devils Hole when he arrived in the area.[18] As the fish Wales collected there served as the basis for the species' naming and ultimately, its conservation, Bradford literally pointed the way to pupfish preservation.

We know about Bradford's attempt to put Devils Hole to use because in Nevada and around the West using water requires a lot of paperwork. Bradford and his partners submitted an "Application to appropriate the public waters of the state of Nevada" on February 14, 1914, sending it to Nevada's state engineer, a powerful position overseeing state water rights. The partners provided information about the source of water; amount they planned to use; purpose to which it would be put; where they planned to use the diverted water; cost of improvements needed to bring the water to beneficial use; and amount of time needed to complete the work. Though not the perfect forerunner to the waiting line at the Department of Motor Vehicles, early-twentieth-century water permits are an example of the growing rigor and bureaucratization of government functions typical of this period.[19]

The state engineer looked over their responses and asked for more information—Bradford and Coleman's application initially applied for "all water" from Devils Hole, but Nevada required that they specify a particular quantity. They settled on 18.8 second-feet,[20] equivalent to 13,610.6 acre-feet per year, a quantity so large it would almost surely have been impossible to withdraw from Devils Hole or put to use. After initial review, applicants were required to complete other tasks for a permit to be maintained: a description of the application needed to be published in a newspaper, and maps and "proofs," typically affidavits, needed to be submitted demonstrating the commencement and completion of work. Permits could also be canceled if deadlines for these tasks were not met—as happened to Bradford's Devils Hole application.

All these rules (and others) stemmed from a novel legal doctrine for water usage born in the West known as the "appropriation doctrine" or the "doctrine of prior appropriation." While elements of the doctrine dated to the 1860s, Nevada codified this water rights system through a series of statutes passed between 1903 and 1913.[21] As articulated by these laws, water rights in the state had three key features. First, water in state belonged to the public, and thus it was a function of state government to ensure that water was applied in the public interest. While the public owned the water, the right to use it, once acquired, could be transferable as other kinds of private property. This made water potentially very valuable to those people who could control it.

Second, to allocate this scarce resource, the appropriation doctrine established a priority system known by the phrase "first in time, first in right." The first user of a river or a spring to put water to beneficial use became a "senior" user; subsequent users became known as "junior." If the available water in a river dropped during a drought, the first, senior, right holders would still be able to withdraw the full amount they were authorized to put to use, while subsequent rights holders might be either unable to withdraw their full allotment or anything at all. Because water sources were relatively rare, and the places where the water was wanted—for a mine or a farm—might be far from a point of origin, it was not necessary to hold land adjacent to a river or stream in order to have a right to use it.[22]

Finally, to gain a right to use water, a person needed to show the ability to put a particular quantity of water to "beneficial use." Nevada's 1913 water law stipulated, "Beneficial use shall be the basis, the measure and the limit of the right to the use of water."[23] This sentence—perhaps *the* guiding principle of water allocation in Nevada—remains on the books to this day. And while the history of water rights law can be confusing and arcane, one of the keys—and continually controversial parts of that law—revolves around a simple question. What exactly counts as a beneficial use?

Almost all those who filed water claims in Ash Meadows during the first half of the twentieth century answered that question by pointing to some readily recognizable beneficial use—mining, agriculture, stock raising—the kind of thing westerners dreamed about. In 1948, however, a conservationist and Stanford University student named Richard Gordon "Dick" Miller—no relation to Carl Hubbs's graduate student and son-in-law, Robert Rush Miller—submitted a curious water permit application to the office of Nevada's state engineer. Miller wanted to "use" water to protect pupfish. The episode put in stark relief how the state's system for water allocation could not incorporate the needs of pupfish, foreshadowing the difficulty facing the Park Service in the 1960s when it attempted to shield Devils Hole from the effects of other water users. Essentially, he found that there was no box to check on state water permits for pupfish.

Dick Miller was thirty-five years old when he submitted his application to the state engineer. He had already earned a master's degree in biology from Cornell University, and after three years as a naval officer on weather ships and minesweepers during World War II, he worked at the Nevada State Museum and as an instructor at the University of Nevada.[24] In Reno, he was a colleague of Ira La Rivers, the professor and conservation vigilante who transplanted and preserved pupfish in vast quantities during the late 1940s and early 1950s. By the fall of 1948, however, Dick Miller had returned to graduate school to earn a doctorate at Stanford, and that December he visited Devils Hole, collecting twenty-eight pupfish from the pool with an improvised net made of wire and a window curtain. Able to keep twenty of the fish alive during his return to Palo Alto, Miller eventually established them in an aquarium in his office at Stanford. (Like others before and after, Miller was unable to rear a second generation in captivity, and he had just two specimens remaining by April, when he stopped keeping notes.)[25]

It is not certain what inspired Dick Miller to file for a water right at Devils Hole. Certainly, his recent trip put the pupfish near the forefront of his thought. Years later, he told Joseph Wales—who, while also a Stanford student, had identified the Devils Hole pupfish as a unique species—that he was worried over highway development in Nevada and the possibility that a new road might "come close to Devils Hole and cause excessive use of the water to heat a gas station or something."[26] But I like to think part of the reason came from his attempt at raising the pupfish in an aquarium. For the first ten days after returning from Devils Hole, and before he moved the fish to Stanford, Miller kept them at his home, literally living with the pupfish. Perhaps this intimacy inspired his effort?

Regardless of the specific motivation, by the end of December, Dick Miller had completed the state engineer's application for a water permit as best he could, writing that he wanted to use the water for the "propagation of fish for scientific and aquarium purposes." Tellingly, he left blank the space on the form for the amount of water to be used, the number of acres to be irrigated, or the industrial use to which it would be put.[27] The water should stay in place for the fish.

This was not the kind of water rights application that the state engineer's office typically received. In correspondence with the water rights surveyor whom Miller contracted to assist with the permit application, Assistant State Engineer Hugh A. Shamberger (and future state engineer) admitted he was "at a loss at the moment as to just how the proofs could be filed so that a water right would be set up."[28] Though interesting, the protection of fish did not fit into the way that water rights were administered. Indeed, nonuse of water could be grounds for permit revocation, not its purpose.

While the principles at stake were very clear to the state engineer's office, Nevada water officials have an ambiguous legacy with respect to

Miller's permit. On one hand, pupfish were invisible to the purposes of the water rights system and thus could not be protected as Dick Miller hoped. But on the other, like John Bradford, who started to build a tunnel to Devils Hole only to play a role in pupfish science and preservation by giving Joseph Wales directions to the site, the state engineer's office also played an unexpected role in the story. Schamberger, the assistant state engineer, suggested that the federal Bureau of Land Management could withdraw the land around Devils Hole from entry to protect it. He even offered to discuss it directly with BLM.[29] And a year later, in the spring of 1950, BLM swooped in as the *deus ex machina* to our tale and proposed to the Park Service that it make Devils Hole a national monument.[30]

The fact that the state water agency suggested federal management of Devils Hole is at first curious. As we will see, in the 1970s Nevada and its state engineer opposed federal efforts to assert a water right at Devils Hole. Schamberger's recommendation, then, can be explained only by the caveat he offered. From his perspective, while the Bureau of Land Management and the National Park Service could set aside Devils Hole and prevent people from appropriating water directly from its pool, these federal agencies still would not have a water right. "If the waters were tapped at some other point and diminished the supply in Devils Hole," Schamberger explained, "there would be nothing that could be done about it."[31] If beneficial use lowered the water table, Devils Hole was still in trouble.

• I Drink Your Milkshake

The hydrologic, or water, cycle is a familiar concept even if the term may not be. Water in the ocean evaporates, forming clouds that produce rain or snow, which feeds rivers that eventually run back into the oceans. Even when we divert water for agricultural or industrial use, this loop of precipitation and evaporation does not change dramatically. A fraction of the water precipitating on land, however, does not return to rivers or oceans. It infiltrates soil and collects in permeable rock, forming underground, waterlogged formations called aquifers. This subsurface water constitutes an important part both of the hydrologic cycle and the human history of water use in the West. Once an aquifer is tapped through drilling and pumping, we call the water brought to the surface groundwater. The scale of these aquifers can be formidable. The Ogallala Aquifer, for example, underlies parts of eight states on the Great Plains of the U.S., and groundwater from this formation irrigates some 15 million acres, making this semiarid landscape one of the primary agricultural regions in the nation.[32]

The Devils Hole pupfish owes its evolutionary history, as well its survival through the twentieth century, to such an underground flow system. After the protection of the land immediately around Devils Hole in 1952, however, the aquifer also became a vulnerability for the pupfish. Snow and rain falling north and east of Devils Hole, including on the Spring Mountains—the range of occasionally snow-capped peaks visible from the Las Vegas strip—slowly infiltrates the soil and rock and then collects in fissures that flow south and west.[33] In Ash Meadows, an impermeable barrier pushes some of that water back to the surface and into the springs that dot the area and make it the largest oasis in the region. Devils Hole, meanwhile, provides what two researchers have called a "skylight to the water table," a geological feature we can almost never, by definition, see.[34] From Ash Meadows, the underground flow continues on to Death Valley, where writer Ed Abbey once darkly speculated that the body of one of the divers who drowned in Devils Hole "may be found someday wedged in one of the outlets of Furnace Creek springs," near park headquarters some 32 miles west.[35] While hyperbolic, the ocean of water underneath the desert creates invisible connections.

Because of the difficulty of actually digging and drawing water from below the ground in quantities enough for widespread irrigation, substantial groundwater development was not an important feature of nineteenth- and early twentieth-century agriculture in the United States. Some of the first widespread use of groundwater for irrigation in the United States came in the west Texas plains, drawing on that massive Ogallala Aquifer. Technical improvements in pumps and drilling techniques, as well as in the availability of motors to power them, allowed for the expansion of groundwater pumping and irrigation beginning in the 1910s, but it only began booming in the period between the 1930s and 1950s.[36] Since then, groundwater has become an essential source for both irrigation and public water supplies. In the United States in 2015, for example, 48 percent of water used for irrigation, and 39 percent of the public water supply, came from groundwater sources.[37]

Surface water has the quality of a renewable resource. While it can be mismanaged, water will, essentially, continue to fall from the sky (even if climate change alters when, where, and how much falls). Groundwater, though, is different. The time scales on which aquifers naturally "recharge" are longer than their current, pumping-induced rate of discharge in many places. This means that pumping can lower the water table, the depth at which the earth is saturated with water. The Ogallala Aquifer is facing a crisis for this very reason. Parts of it have declined by more than 150 feet, leaving some farms and towns high and dry.[38] In Nevada, by the 1950s, groundwater pumping around Las Vegas (Spanish for "the meadows") removed so much water from the aquifer beneath the valley that it not only dried up its springs and led to the extinction the valley's native fish but also caused the valley itself to subside.[39]

The ability of groundwater pumping to lower the water table and reduce spring flow elsewhere in a basin brings to mind the climactic, murderous finale to the 2007 film *There Will Be Blood*. Writer and director Paul Thomas Anderson's meditation on greed, faith, and family, the film also includes some truth about oil fields: viscous, yet liquid, oil does not stay in one spot (like iron ore) but rather exists in large pools that can move across property boundaries. While the water cycle is very different from the forces that create oil deposits, pumping oil or water in one place can reduce the possibilities for pumping it somewhere else. In the film, radio evangelist and minister Eli Sunday (Paul Dano) cannot understand why California oilman Daniel Plainview (Daniel Day-Lewis) refuses to lease land that Sunday stands to profit from, even though it is located at the center of Plainview's sprawling oilfield. Plainview drunkenly explains that he does not need the land since he has wells all around it, sucking up the same oil. "I drink your milkshake!" he screams, "I drink it up!" He then bludgeons Sunday to death with a bowling pin.

Pumping water from one area in an aquifer can also reduce its level nearby. In Ash Meadows, if someone developed enough groundwater pumps, they could lower the water level in Devils Hole without ever putting a straw in the pool itself. Bradford's project would have drained Devils Hole by turning it into Ash Meadows' water tower, but widespread groundwater pumping elsewhere in Ash Meadows could also lower the water level in Devils Hole. The effect would be the same: I drink your milkshake.

Into the 1950s, when Death Valley took over management at Devils Hole, the potential effect of groundwater pumping on the pool was academic. Across the entire Amargosa Desert, of which Ash Meadows is a part, the U.S. Bureau of Reclamation could later only uncover evidence of eight wells being drilled before 1954.[40] (Specifically within Ash Meadows, state well drilling records show no wells drilled before 1960.[41]) As far as agriculture went—irrigated by either surface water or groundwater—Ash Meadows and the Amargosa remained a backwater. Still, change was underfoot. Nevada had updated its water law in 1939, giving the state engineer's office as much authority over the management of groundwater as it already had with surface waters. Afterwards, applications to appropriate groundwater poured in. In 1940, less than half of all water applications filed in the state were for groundwater. Just fifteen years later, however, 85 percent were for groundwater.[42]

The explosion in groundwater pumping across Nevada, as in much of the West, has resulted in a massive problem with "overappropriated" groundwater basins. In theory, each separate basin should be managed such that rights to withdraw water from a basin equal the annual rate of recharge from precipitation so that the use of water is sustainable. Over decades, however, successive state engineers have allowed for the approval of more water rights than there is available water. By 2018, of the 256 basins in Nevada, 84 were out of balance and 26 had water rights for 300 percent more water than is sustainable.[43] One of these basins is right next door to the Amargosa Desert basin of which Devils Hole is a part. In the Pahrump groundwater basin, the sustainable "perennial yield" is estimated to be 20,000 acre-feet, yet there are more than 59,000 acre-feet of rights. This does not include water removed from the basin's 11,000 domestic wells.[44]

In the 1950s, of course, these problems lay in the future. The first person to imagine putting groundwater to "beneficial use" in Ash Meadows was a man named George Swink. What drew Swink to Ash Meadows is uncertain, but he first contacted the Division of Water Resources about area water rights as early as 1958, when he told Hugh Schamberger he had obtained an option on more than 3,500 acres in Ash Meadows.[45] (In a 1957 administrative reshuffling, the legislature made the state engineer the chief of a new state agency, the Division of Water Resources.) By 1965, Swink claimed to have invested $1.5 million in developing the area.[46]

Born in Colorado in 1896, Swink's family had pioneered cantaloupe and watermelon growing along the Arkansas River southeast of Colorado Springs during the last decades of the nineteenth century. As a young man, Swink found his way to California. In 1917, he was working at an auto supply shop in the Imperial Valley, a booming desert agricultural region in the state's far southeastern corner.[47] By the 1930s, he had become a prominent melon grower in the region.

Among the things that Swink likely learned there—aside from playing a role in suppressing a farm workers' movement, a bloody episode chronicled in the 1930s by journalist Carey McWilliams's exposé, *Factories in the Field*—was substantial experience in irrigated agriculture.[48] By the 1960s, he was no longer living full time in the valley but two blocks from the beach in La Jolla, the wealthy enclave north of San Diego. His home was less than a mile south of the iconic pier at the Scripps Institution, where Carl Hubbs worked.[49]

Swink had two very different visions for the Ash Meadows landscape, both of which he pursued during the early 1960s through his firm, the Nye County Land Company. At first, he imagined building a large agricultural and ranching operation, the likes of which were already familiar to him from his experience in California. "[B]y the fall of 1961 I had determined that the area had tremendious [sic] potential, there is perhaps 5,000 acres of as fine land as lays out doors, that will grow anything we grow in this entire Southwestern country, and with an abundance of fine water it has an unlimited possibility," Swink later wrote.[50]

Around the same time, however, he discovered that the Atomic Energy Commission was considering building a bedroom community for workers involved with its nuclear rocket development efforts at its massive 1,300-square-mile Nevada Test Site. The Test Site's closest border is just 20 miles north of Ash Meadows. In the summer of 1963, bills were introduced into the House and Senate to devote $11 million to acquiring land and constructing a new town.[51] At hearings held in the fall, Ash Meadows was named one of the potential sites for the community.[52] With all of Swink's water rights filings and landholdings, the federal government would have to compensate him handsomely if it moved forward with the town. By January 1964, however, the funding for the town was called into question in light of a "possible cutback in Government support for the nuclear rocket program," according to a report by the Joint Committee on Atomic Energy. The bill, and the community, were shelved.[53] Perhaps the drawbacks of airborne nuclear engines became apparent to congressional leaders, but the decision seems to have been about changing priorities within President Johnson's military budget more than anything else.[54] Swink later lamented that if President Kennedy had not been shot, "That city would have been built by now, but that is another story and one we have no control over."[55]

After the federal government's interest waned, Swink continued to consider developing a city, taking on a partner and receiving some publicity, including an Associated Press story in February 1964, in which Swink claimed he would "put up 800 homes, a school and public commercial buildings beginning in May," with occupancy beginning by the start of the next school year. He even came up with a name for the development, Lakeview, apparently for the chain of lakes he envisioned in Ash Meadows.[56] The partner Swink took on for the project became ill and left him in the lurch. The development stalled, and Swink returned to the agriculture idea.

Whether for a nuclear community or for the production of agricultural commodities, however, Swink needed water. Between 1961 and 1966, he filed applications for over eighty underground water permits. By his own admission, Swink submitted applications for "every possible location for underground water in the area." By doing so, he wrote to the Division of Water Resources, "I beleave [*sic*] I am in possition [*sic*], under your water laws to recover any and all water there may be in the area and at the same time to be protected from claims on the part of other landowners that I am taking their water."[57]

While Swink had big plans for bringing water to beneficial use, his project inched forward at a glacial pace on the ground. Transforming Ash Meadows into a large, productive agricultural area—or a town—may have been too much for him. In 1963, the year he turned sixty-seven, newspaper reports suggested he hit and killed two hitchhikers whose car had broken down along Highway 95 outside Needles, California. (This event, in addition to being emotionally trying, also kept him in court.[58]) He did build some fencing for cattle and began impounding some water. But despite all his water applications, he had permits in danger of cancellation because of failure to begin the work of bringing that water to beneficial use. By February 1966, he had drilled just two wells.[59]

The Park Service might not have known of Swink's dream for Ash Meadows during the first half of the 1960s if not for a newspaper article. In November 1962, the *Las Vegas Sun* ran a story penned by an aggrieved homesteader who resented a large developer buying up properties and filing water rights in Ash Meadows. The author noted that a "large-scale real estate operator from California"—Swink—had gone so far as to file for a water right at

Devils Hole.[60] Indeed, that year, among his flurry of groundwater applications, George Swink proposed to pump water out of the Devils Hole pool and then send it downhill in pipes to irrigate 1,560 acres in the northern portion of Ash Meadows.[61] Cheap gasoline or electric pumps could do the work that Bradford's abandoned hand-dug tunnel would have. Bradford, who died in 1931, might have been jealous.[62]

Death Valley National Monument had not been an attentive steward of Devils Hole—it fought against acquisition and then did little to incorporate the site into the monument, neither learning about the ecosystem nor offering much in the way of visitor interpretation. Still, Monument staff reached out to NPS's own water resources branch in Washington, D.C., to let it know about Swink's water permit.[63] In reply, the agency's water resources branch admitted it knew "very little about the topography, hydrology, and geology of the Devils Hole area."[64]

Confronted with its own ignorance on Devils Hole, Death Valley staff then asked the U.S. Geological Survey to investigate. Somewhat contrary to its name, this Interior Department sister agency had been interested in water since the days of John Wesley Powell in the late nineteenth century. USGS Carson City branch chief George F. Worts Jr. canvassed what was known of Devils Hole hydrology and delivered a report in August 1963 that put NPS at ease. Read carefully, however, Worts's conclusions offered a prophetic warning. Pumping directly from other Ash Meadows springs could affect the pool level in Devils Hole, and "if deep wells were drilled," Worts wrote, "pumpage a mile from Devils Hole probably would affect the pool level within a year."[65]

The Park Service, however, did not consider this likely. After reviewing the report, E. W. Reed, the new chief of NPS's Branch of Water Resources, wrote, "There appears to be no immediate danger to the pool unless wells are drilled into the carbonate rocks close to the pool or the springs to the west are pumped or otherwise developed extensively so as to lower the head at the springs. Both of these possibilities seem remote at the present."[66] By the end of the decade, that "remote" possibility would be well under way.

The Park Service Branch of Water Resources also assured the Death Valley staff that it was unlikely that the state engineer would approve Swink's permit to pump directly from Devils Hole. On this count, the Park Service water administrators were correct. Indeed, the state engineer's office had proposed in the 1940s that the federal government reserve the land around

Devils Hole specifically to prevent this kind of permit from being approved—even though it did not recognize the pupfish's or the federal government's right to the water should it be sucked out through groundwater pumping.[67]

Despite Swink's big plans for Ash Meadows, most of the transformations he engendered took place on paper—filing water rights and buying property. In the mid-1960s, Ash Meadows itself looked much as it had a decade earlier: a few people living there, some cattle grazing, and not much else. The most exciting happening may have come from the corner of the bar at the Ash Meadows Rancho, a brothel and watering hole on the southern edge of Ash Meadows, where Ed Abbey was putting the finishing touches on his classic book, *Desert Solitaire*.[68] As late as 1967, the threat of groundwater pumping was not obvious, even to those familiar with Ash Meadows. That summer, Robert Rush Miller, more than twenty years removed from his dissertation on the pupfish, revisited Ash Meadows and prepared a report for NPS in which he argued that the "principal danger to the Devils Hole pupfish" was that "some 'crank' or irresponsible person might try to wipe it out."[69] Groundwater pumping went unmentioned.

Highest and Best Use

After years of work on developing Ash Meadows, but with little to show for the effort, George Swink called it quits. In 1966, he sold his land holdings and water filings. He was done wrestling with the state over water-permit extensions, done with vacillating on what exactly the goal of the development should be, and done trying to put Ash Meadows water to beneficial use. "I only wish I was about twenty years younger,...but as it is, I have one hell of a time keeping on my feet," he lamented to a Nevada Division of Water Resources official.[70]

The purchaser of Swink's water permits and 5,000-odd acres of land did not have this problem. Twenty-three years Swink's junior, Francis Cappaert spent the next three years expanding the scope of the Ash Meadows project—including engineering a large federal land deal—and began reshaping the landscape, with dramatic consequences for the Devils Hole pupfish. The invisibility of the pupfish to Swink, John Bradford, and other residents in Ash Meadows had not doomed it to extinction, mostly because those dreamers were unable to bring their ideas for intensive irrigated agriculture to fruition. But once Cappaert began building what Swink had imagined, the invisibility of the pupfish to the water rights system would be a serious impediment to the species' survival.

Born in 1919 and raised in Michigan, Cappaert served with distinction on a PT boat during World War II, earning a Silver Star for his service in the Pacific Theater.[71] Moving to Mississippi, Cappaert started a mobile-home manufacturing business, which he grew into a goliath. Months after buying up Swink's property, he became the chief executive and majority owner of Guerdon Industries, which *Forbes* magazine described as "probably… the largest U.S. producer of mobile homes," with sales north of $80 million a year.[72]

Called "just a down to earth S.O.B. who loves to wheel and deal," by Jimmy "the Greek" Snyder, his one-time publicist and noted Las Vegas bookmaker, Cappaert did not set his sights on just leading the mobile-home business.[73] By the early 1970s, he owned upward of 100 companies, according to the business magazine *Dun's*. Cappaert claimed he was worth about $100 million. With bravado, he told *Dun's* he planned to double this figure in five years. Asked how he hoped to do that, Cappaert replied: "Look, this morning I've already made five or six million bucks, and it's not even 10 o'clock. Right now, I'm hanging around my office to close two or three more deals. A good day's work, right?"[74] His investment in Ash Meadows was part of this sprawling money-making machine.

Cappaert kept tabs on each of these enterprises—in sectors from electronics and oil drilling, to manufacturing and agriculture—by flying around the country in a private jet, which served as a very upscale mobile home, though substantially more exclusive than the ones his company built.[75] Despite reportedly traveling some 5,000 miles per week visiting his interests, Cappaert was apparently more absentee landlord than micromanager.[76] In *Dun's*, he said that his managers were free to run their companies as they saw fit, as long as they earned a 10 percent profit. Below that, Cappaert became much more interested, and when profits dipped below 5 percent, he found a new manager.[77]

Learning about Cappaert and his motives is difficult. Despite the fact that his name is on a U.S. Supreme Court decision related to the pupfish, I have been unable to find out some basic facts about his tenure in Ash Meadows, such as how he became aware of Swink's operation or whether (or how frequently) he visited Ash Meadows.[78] After years on this project, I can only take comfort in the fact that Cappaert, who died in 1996, cultivated this secrecy. One of the few profiles about him made a comparison to another eccentric and secretive entrepreneur, Howard Hughes.[79]

When Cappaert's name did find its way into the news it was not always in the most flattering of contexts. He owned Diamond Reo, primarily a manufacturer of trucks for the U.S. military, and ran the company into the ground in less than five years. Though the company was already stressed prior to his acquisition, "Cappaert" became a dirty word among former Reo workers in Michigan who lost their pensions in the company's collapse.[80] Released U.S. State Department cables from 1973 and 1974, meanwhile, described a convoluted situation wherein a debt-laden, Cappaert-owned, air-transportation company in Costa Rica surreptitiously flew a DC-6 airplane to Cappaert's adopted hometown of Vicksburg, Mississippi, without the knowledge of Costa Rican creditors, who apparently worried they would never see the plane again.[81] Most publicly, Cappaert found his name wrapped up in the fallout from Watergate—with the *Washington Post* identifying him as having donated $210,000 to Richard Nixon's 1972 reelection campaign and federal agents later investigating the contribution's legality.[82]

Whatever his motives and methods, and however he even learned about this corner of the desert, the secretive and charismatic Francis Cappaert brought substantial new energy and capital to Ash Meadows. He also brought a new idea: instead of just putting water to beneficial use on the patchwork of private land Swink had managed to acquire in the area, Cappaert's company, initially called Spring Meadows Inc., but later renamed Cappaert Enterprises, could expand his landholdings by acquiring vast tracts of federal land, including, he hoped, land adjacent to the NPS Devils Hole property. "This man Cappaert has great plans for the project, and is going about things in a real manner," Swink believed.[83]

One of the curious situations that makes putting water to "beneficial use"—or, for that matter, conserving water for species—difficult in the American West is that the authority over land use and water use is often controlled by different levels of government. As we have seen, water management is generally held to be the provenance of the states. In Nevada, the Division of Water Resources fulfilled this responsibility. Land policy, by contrast, is often set by the federal government. Despite this important institutional divide, in the 1960s, Nevada's Division of Water Resources and the federal government's Bureau of Land Management shared a critical ideological orientation. Wildlife, including pupfish, did not easily figure into their understandings of the value of water or land. Using the real estate terminology of "highest and best use" to determine the value of its Ash Meadows holdings, the Bureau of Land Management looked on its land and concurred with Cappaert: it was ripe for irrigated agriculture.

Nevada is overwhelmingly public land, with over 80 percent of the state's area owned by the federal government—the highest percentage in the lower forty-eight states.[84] These public lands are predominantly administered by the Department of the Interior, a rather misnamed agency because, as historian Megan Black has observed, Interior's role for several decades after its creation in 1849 was to incorporate the "exterior"—lands recently won through the dispossession of Native Americans and European colonial powers.[85] In the case of southern Nevada, this land only became public in the eyes of the federal government in 1848 after the signing of the Treaty of Guadalupe Hidalgo, which ended the Mexican-American War and added much of the present-day Southwest to the United States. Public lands are the spoils of war.

Having acquired all this land, the main question during the nineteenth and early twentieth centuries was how to get rid of it as quickly as possible. With the exception of military reservations, and later national forests, parks, and monuments, Congress wrote laws that encouraged individuals to develop the land—the homestead of John Bradford, whose aborted tunnel to Devils Hole is still visible on the desert floor, was a testament to this approach. Other laws, aimed at ranchers, miners, and logging companies, preceded and followed the 1862 Homestead Act. All passed with the notion that the land would be put to best use in private hands.

That Nevada is still overwhelmingly public land might suggest that the privatizing impulse in federal policy failed. One of the reasons for this—as this chapter has already explored—is that aridity made putting the land to use difficult without adequate water. Starting in the 1930s, when it recognized that selling off the public domain might not only be impossible, but undesirable, Congress reversed course, pursuing new laws that sought to manage public lands in the public interest. The Bureau of Land Management, established in 1946, became one of the main levers for this new policy. BLM received further direction from Congress in the form of the Classification and Multiple Use Act of 1964, which forced the agency to publicly define and prioritize the values for its tracts of land all across the West, including in Nye County, Nevada. Much remained the same: lands outside of parks, monuments, and military reservations could still be *used*, and BLM continued to award contracts for timber, grazing, and mining. Yet, the agency's implementation of the Act meant that much land was closed to privatization for agricultural uses through the Homestead Act and other nineteenth-century land laws and that additional acreage was protected for watershed health, recreation, and wildlife.[86] For Cappaert, this shift in federal policy from disposal to management meant that it would be much harder to actually acquire cheap, public-domain land for agriculture than it had been just a few decades earlier.

The law, though, tends to be full of helpful exceptions for those who know how to find them. Section 8 of the 1934 Taylor Grazing Act held one such caveat. The act created a leasing system for owners of cattle wishing to graze livestock on federal property, with management of the "grazing districts" conducted by the new Division of Grazing, a forerunner to BLM. Section 8 supported this purpose by allowing for land exchanges between private land owners and the federal government in cases where doing so would improve the management of public lands grazing districts, or, in the words of the act, the "public interests will be benefited thereby."[87] In an ideal case, Section 8 would allow for the federal government to acquire private land in the midst of a larger grazing district, while offering the private owner land from the public domain of equal value that was not needed for this management program.

Each BLM state office in the west directed its own exchange program and undertook exchanges at its discretion. As a result, the amount of land swapped between the federal government and private landowners varied wildly from state to state. During the ten years between 1959 and 1968, Oregon and Nevada each exchanged over 250,000 acres, while Wyoming and California each exchanged under 10,000 acres.[88]

Perhaps the most notorious Section 8 land exchange in Nevada history involved Howard Hughes, who in the early 1950s acquired 25,000 acres for a planned research lab for Hughes Aircraft just west of the Las Vegas strip. Though a field examiner in the Interior Department believed the exchange "would certainly not be to the advantage of the government or the general public," Hughes managed to push the deal through by drawing on his lawyers and military contacts.[89] (The land was never used for the research facility and sat vacant for nearly forty years before being developed into the upscale Las Vegas suburb of Summerlin.) The memory of this land deal even influenced the media's framing of Cappaert's own land exchange. After it became public, the press wire service UPI wrote a short story, picked up in at least one Nevada newspaper, that told readers that the "industrialist Howard Hughes acquired most of his 40-square mile 'Husite' near Las Vegas under terms of a similar exchange."[90] Perhaps Cappaert, the Hughes-watcher, was proud.

Cappaert began formulating a plan to make use of Section 8 from the moment he acquired Swink's operation. The company offered the federal government private land in northern Nevada, that, according to the Bureau of Land Management, would add additional grazing land, create more contiguous parcels of federal land, and could be valuable for recreation purposes.[91] In return, between September 1966 and March 1967, the company submitted eight different applications for public land in Ash Meadows—totaling more than 13,000 acres. Added to the 5,000 acres he already bought from Swink, Cappaert stood to control the better part of Ash Meadows.

Cappaert's firm made two connections that helped move his land exchange forward but that could confirm—for the cynic—that land exchanges were more often for the rich and powerful and not in the public interest. First, Spring Meadows Inc. contracted with a retired associate director of the BLM, Harold Hochmuch, to navigate the bureaucracy and "expedite matters" with BLM, also making him an officer of the company.[92] Second, the company hired a Reno-based law firm with two important founders. One partner in the firm, Robert McDonald, became a director of Spring Meadows Inc. and provided early legal counsel to the company.[93] The other partner was none other than the U.S. senator from Nevada, Alan Bible. At the time, under the chamber's rules, senators were able to maintain private law practices while in office, and while Bible apparently had not practiced since 1954, his name stayed attached to the firm.[94] Bible's connection to McDonald and the law firm went beyond letterhead: Bible received contributions to his senate campaigns in 1956 and 1962 from McDonald, and by 1970 Bible's son, Paul, worked at the firm.[95]

Cappaert's link to a sitting U.S. senator did not go unnoticed by BLM personnel working on the land exchange: an official in the agency who worked on the deal later told another federal employee that the "Cappaert party had initiated and pressed for the exchange, obviously under the political influence of Senator Bible of Nevada." The official also claimed he was "invited to lunch" to discuss the exchange with Senator Bible's son and Cappaert's attorney who were "evidently... dear friends."[96]

Investigating the origins of the land exchange, I have uncovered no obviously illegal activity on the part of Bible, Cappaert, or anyone else, but tracing the stench of influence pedaling and impropriety surrounding it was deeply satisfying. It got me close to something we often think we know, or at least suspect, but cannot always well document: that wealth and personal connections shape our politics to the detriment of the commonwealth. Combined with the other information I could pull out from Cappaert's business career during the same era, it also makes it tempting to paint Cappaert as a real villain in the pupfish story.

But leaving the story there misses the more significant feature of the land exchange: at no point during its consideration did the public officials at BLM address its effect on wildlife. Whether the exchange moved forward or not may have depended on whom one knew or whom one could hire to navigate a complex bureaucracy, but the process at no point investigated what it would mean for pupfish. In land exchanges, as in water permitting, pupfish were invisible.

This point was distilled nicely in a thin, two-page "land examination report," composed by BLM and buried in the middle of the files related to the land exchange at the National Archives near San Francisco. Despite the difficulties of farming in the Amargosa Desert, an effort that would require a "great investment to reclaim this land through leveling, soil additives and proper irrigation," BLM concluded that the "highest and best use" for the land was agriculture.[97]

• One Mile

In July 1967, BLM announced its plans to classify the 13,663.19 acres in Ash Meadows that had been selected by Spring Meadows Inc. as available for "disposal."[98] In BLM-speak this meant that the land could be exchanged. Under agency rules, however, it had to offer sixty days for any protest to come forward before publishing a final decision.

Though the plans had been underway for nearly a year, publication in the *Federal Register* represented the first time that NPS had heard about it. This is perhaps unsurprising, given that the agency had not been an attentive steward of Devils Hole during the site's first fifteen years of management. The notice of the exchange, however, represented a major shift in the Park Service's attitude. Staff at Death Valley had initially opposed the acquisition of Devils Hole, and despite the conclusions of the 1963 USGS-Park Service report—that groundwater pumping near Devils Hole could cause the pool level to decline—NPS believed the potential for such water development was "remote."

After the *Federal Register* ran the details of the BLM-Cappaert exchange, the scales fell from Park Service administrators' eyes. In addition to the sheer size of the exchange, some of the land selected by Cappaert directly abutted the NPS Devils Hole property. Putting the pieces together, Death Valley superintendent John Stratton realized that development in the area would "increase water use which will in turn lower the water level of Devils Hole and adversely affect the environment in which these small fish live."[99] He asked for additional time to investigate and develop a reply to BLM.

And just like that, the people and institutions that cared about pupfish and the people and institutions that did not notice pupfish collided. As in the previous moments where these groups ran close to each other—in 1914 when Bradford began digging his aborted tunnel, or in 1948 when Richard G. Miller filed a water right on Devils Hole in order to protect the pupfish—the tension did not center on dislike or animus toward the fish. Instead, it was a matter of seeing. The imperative to use, to build, to grow simply did not notice that water and land were also home to pupfish. Until 1967, the pupfish persisted despite this impulse toward development, but beginning with the BLM land classification marking the land as disposable, the survival—or extinction—of the pupfish ran directly through institutions that did not notice pupfish.

This did not bode well for the pupfish for several reasons. First, the Park Service itself did not have much information to support a protest. Its best defense against the exchange was the 1963 USGS report it had commissioned after George Swink filed for a water right on Devils Hole's surface water. But the conclusions of that report did not give any definitive guidance on how groundwater pumping would affect the pool level, and the Park Service had taken no further action to build a more detailed portrait of the aquifer beneath Ash Meadows. Reed, the NPS Branch of Water Resources chief who viewed groundwater development as a "remote" possibility in 1963, now said that while "some form of protest" to the BLM exchange was warranted, "the information available here is not sufficient to define a clear course of action."[100]

Second, all the Park Service could do was protest a decision to privatize more land in Ash Meadows. It could not know exactly what Cappaert's plan was, nor would protesting the land classification prevent water from being pumped, since BLM did not control water rights. If "we could be assured that the lands surrounding Devils Hole would continue to be used only for stock raising," Reed wrote, "our concerns for the preservation of this area would vanish. However, with the bulk of the lands in private ownership we would have no such assurance."[101]

Finally, and most importantly, the exchange took place before the passage of modern environmental laws. In the era after the National Environmental Policy Act (1969) and the Endangered Species Act (1973), the burden of proof would be on BLM to show that the exchange *would not* harm the Devils Hole pupfish or other Ash Meadows endemics. But in 1967, NPS found itself in the position of having to marshal evidence to submit to BLM to show that the exchange *would* harm the species. And BLM was not necessarily predisposed to accept that NPS was allowed to petition it. Before ultimately agreeing with the NPS request to withdraw some acreage from the exchange, one BLM official thought that "perhaps the Park Service lacks legal authority" to protest BLM classifications.[102]

Facing these constraints, in the fall of 1967 NPS ultimately requested that all lands selected by Cappaert within a mile of Devils Hole be removed from the exchange. The one-mile figure was used in the 1963 USGS report. There, it served as a ballpark figure, but with little else known about the area's hydrology, the one-mile distance provided the only justification NPS could give for opposing the exchange. Out of the more than 13,000 acres included in the planned exchange, it protested 560.[103] Six months after receiving the Park Service protest, BLM released its final classification, accommodating NPS's objection while offering its sister agency a disclaimer: "the Bureau of Land Management administers the public domain adjoining Devils Hole but the control and issuance of well permits for groundwater is the authority vested in the Nevada State Engineer's Office."[104] Whatever ended up happening to the Devils Hole pupfish, or to the land it had just agreed to hand over to a large ranching corporation, was not BLM's responsibility, it seemed.

As the paper transformations of Ash Meadows wound their way through BLM offices, Cappaert's company did not wait around to begin reshaping Ash Meadows. In October 1966, nearly a year before NPS even became aware of the exchange, a driller punched a 248-foot well into the earth just 900 feet from Devils Hole (fig. 3.1).[105] It was just one of the 13 wells that the company had in place by 1967. Consulting engineers for Spring Meadows tested the well near Devils Hole and reported pumping it at the rate of 900 gallons per minute for more than two-and-a-half hours.[106] "Although no decline in water level was registered on the recorder at Devils Hole... it is probable that extended pumping of this well would ultimately be reflected in a lowering of the water level in Devils Hole," the engineers explained in a report to Cappaert.[107] Based on data from the test, the hydrologists calculated that if the well were put into regular use for 100 days the water level in Devils Hole would decline by 2.2 feet.[108] Thus, in June 1967—before word of the exchange became public—Spring Meadows Inc. had the critical information in hand to know what pumping in Ash Meadows might do to Devils Hole. But legally it did not need to do a thing.

Moreover, Cappaert's contractors had a much better idea of what would happen at Devils Hole in the event of pumping than NPS. When Death Valley staff stumbled on the well in fall 1967, they were stunned: "It is apparent that the potential damage... to Devils Hole water level," one employee wrote, "is already in existence."[109]

CHAPTER 4

Save the Pupfish

On a warm spring day in 1969, a group of federal and state employees left Death Valley National Monument headquarters at Furnace Creek and caravanned up the grade and out of Death Valley to the southeast. Two dusty sedans and a station wagon crested the 3,000-foot pass and descended the eastern side of the Funeral Mountains into the Amargosa Desert. The party constituted a veritable Who's Who from the region's public scientific and land management bureaucracies, with staff representing three different federal agencies—NPS, BLM, and the Bureau of Sport Fisheries and Wildlife—as well as California's and Nevada's fish and game departments and the University of Nevada, Las Vegas (UNLV).[1]

Arriving in Ash Meadows after the 50-mile ride, the group found the landscape transformed. While the land exchange between BLM and Spring Meadows Inc. would not be finalized for several months, Cappaert's company was already hard at work remaking the property it had purchased from George Swink. Near Point of Rocks, a group of springs and seeps about 2 miles southeast of Devils Hole, the party observed hundreds of acres in the process of being leveled by heavy equipment. They also saw a pump perched over one of the springs, drawing down the water level and threatening its population of Ash Meadows Amargosa pupfish. And near Devils Hole they stood next to the well Cappaert's firm drilled in 1966 and that the Park Service first observed the following year.[2] It appeared to have recently been equipped with an electric pump, though it had not been put into operation. "Some concern was expressed," a BLM fishery biologist noted in the muted style of bureaucratic memoranda, "that full pumping could possibly reduce the level of Devils Hole."[3]

They saw more than just pumps, irrigation ditches, dams, and heavy machinery. The scene before them represented the culmination of dreams stretching back to John Bradford's aborted tunnel to Devils Hole. In contrast to those previous visions of development, which went unrealized because its dreamers were too poor, Ash Meadows too remote, or the benefits too uncertain, this time it seemed as if the possibility of putting Ash Meadows' water to use would be brought to fruition. Visiting the altered habitats of Ash Meadows, the caravan grasped just how invisible pupfish and other endemic species were before the unfolding logic of beneficial use.

Part of a three-day, interagency tour of desert fish habitats, the April 1969 trip also clearly revealed the Park Service's 1967 protest to the BLM-Cappaert exchange for what it was: a joke, an accommodation incapable of preventing harm to Devils Hole, never mind the other springs and species around the area. Cappaert's firm could live without the 560 acres the Park Service wrested from the planned exchange and go on farming thousands of others. Even after the field trip, when BLM staff realized that the well at the corner of the Devils Hole property had been drilled on federal land without a permit, Cappaert soon had more than a dozen other legal wells in the area ready to pump the same water.[4] Indeed, the water level in Devils Hole had already started plummeting by late 1968 without the well closest to it in operation.[5]

If Cappaert's agricultural development continued—as they had every reason to expect—biologists imagined a disastrous ending not just for the Devils Hole pupfish but also for the three other living fish species in Ash Meadows.[6] The Bureau of Sport Fisheries and Wildlife (BSFW), the arm of the U.S. Fish and Wildlife Service that focused on noncommercial wildlife issues from the mid-1950s until reabsorbed by its parent service in 1974, had an employee in attendance. Clinton Lostetter, the Bureau's point person on endangered species, told his superiors after the field trip that if Cappaert continued pumping the wells around Ash Meadows, "it's goodby[e] pupfish."[7] The "only real uncertainty," a group of scientists and managers explained later, "is just when this will occur."[8] Confronted by a system that rendered the species invisible, the pupfish were goners.

But this did not come to pass. Improbably, over the nine years following the visit by biologists and managers, the pupfish not only managed to survive, but Cappaert and the dreamers of beneficial use lost their right to pump water to the extent that it infringed on the pupfish's habitat needs. The water (mostly) stayed in the ground for miles and miles around the habitat, to the great chagrin of Cappaert, the Nevada Division of Water Resources, and many citizens across Nevada. The same tangle of institutions and individuals—private ranchers, state water agencies, consulting hydrologists, federal land managers—that engineered ever-more elaborate

FIGURE 4.1. Professor James Deacon (UNLV) in front of Devils Hole with artificial lights, artificial shelf, and water monitoring equipment, circa 1974. Photograph by and courtesy of Phil Pister.

ways to develop groundwater across the Great Basin, crashed onto Devils Hole like a wave on a jetty. The pupfish would not be moved.

This chapter and the next explore how the Devils Hole pupfish survived Cappaert's company. The center of the drama was not, however, in the fight between the company and the federal government—though the Supreme Court case, *Cappaert v. U.S.* would suggest as much. Instead, throughout the late 1960s and the 1970s the most difficult questions about whether the pupfish would survive turned on decisions within—and fights between—federal and state agencies themselves. Cappaert's company wanted to use the water in Ash Meadows; that much seemed clear. But how the federal and state governments responded was not foreordained. The participants in the April 1969 tour and their allies understood this clearly. They stared down not only the disheartening prospect of the Devils Hole pupfish's extinction at the hands of groundwater pumping by a private company but also faced a much more disturbing nightmare: leaders at the highest level of the federal government's land management and conservation agencies, along with Nevada's water bureaucracy, might let it happen.

Focusing on whether, and then how, the federal government, in particular, attempted to protect the Devils Hole pupfish serves a number of purposes. First, it provides insight into how large bureaucracies function. We are used to thinking about public agencies as monolithic, but one of the key ways they operate is by variously resisting and accommodating demands from their own employees, from the public, and from other segments of the bureaucracy. This was true when Carl Hubbs criticized the Park Service in 1950, and it remained true at the start of the 1970s.

Examining the tensions within and between public land management agencies also sheds light on a transitional moment for federal bureaucracies' relationship to endangered species. When the biologists showed up in Ash Meadows in 1969, species saving was very nearly uncharted territory at the Interior Department. A decade earlier it is unlikely that such a group of employees would have even assembled at Ash Meadows or posed questions about how to "save the pupfish." As late as 1962, BSFW, one of the agencies represented at the April 1969 Ash Meadows tour, actively engaged in the eradication of native fishes—a practice made infamous in its elaborate effort that year to improve conditions for sport fishes by poisoning Colorado pikeminnow and bonytail chub with a piscicide called rotenone throughout a 450-mile stretch of the Green River in Wyoming and Utah.[9] In Ash Meadows, NPS and BLM also each had checkered pasts with respect to the pupfish.

Fast forward to the latter part of the 1970s, however, and the situation was quite different, closely resembling the institutional arrangement we are familiar with today. After the passage of the Endangered Species Act in December 1973, perhaps the most far-reaching wildlife law in the world, FWS began developing a plethora of rules and procedures for defining how federal agencies and private firms should deal with endangered species and their habitats. As Cappaert's firm began remaking Ash Meadows, however, the passage of that law was still a long four years away; each Devils Hole pupfish only lives for about a year.

And finally, examining how the federal government responded to Cappaert's ranch explains how the pupfish persisted: through an ad hoc, uncertain process. As Cappaert's firm expanded in Ash Meadows, the Devils Hole pupfish's supporters faced a lack of clear responsibilities, limited statutory power, and skepticism toward efforts for protecting the species at the highest levels of the federal government. Field employees and university scientists forced the hands of agency leaders and kept the pupfish alive in memos, reports, and the press until those leaders conceded, making the Devils Hole pupfish one of the first objects of the Interior Department's nascent endangered species program. From this ad hoc arrangement, managers made desperate efforts to save the pupfish, moving fish to new habitats, trying to raise them in aquaria, installing elaborate light fixtures and platforms in Devils Hole, and producing reams of memos charting their successes and failures.

These efforts led federal agencies headlong not only into Cappaert's company but also into Nevada's system for apportioning water, which did not imagine pupfish as a "beneficial use." Challenging this system first required making the species visible and important within federal agencies. To begin with, then, the Devils Hole pupfish persisted through the 1970s because a group of interested government employees arrived in Ash Meadows in April 1969 and wondered what they could do.

• The First Seventy-Eight

The *Federal Register* is the most important publication that none of us ever read. Published five days a week since the 1930s, it compiles all proposals, decisions, and changes to the federal government's regulations and rules as well as presidential orders and proclamations. On a day in summer 2018, it included the announcement of a U.S. Coast Guard plan to alter the schedule of drawbridge operations in the Wilmington, North Carolina, area to accommodate a local triathlon race; a notice for the next meeting of the Nuclear Regulatory Commission; and an announcement of a new service agreement for the U.S. Postal Service.[10] Essentially, the *Register* is a daily guide to what government is, how it works, and what it values.

Endangered species made their initial appearance in the *Federal Register* on March 11, 1967. On page 4001 (pages are counted sequentially from the start of the calendar year), Secretary of the Interior Stewart Udall announced that he made the first additions to a new U.S. endangered species list. "After consulting the States, interested organizations, and individual scientists," the notice read, seventy-eight animals "threatened with extinction" would be placed on the list, including iconic creatures such as the American alligator, the grizzly bear, and the California condor, as well as the ivory-billed woodpecker, which had not been seen since the 1940s. It also named twenty-two fishes, including the Devils Hole pupfish.[11] The pupfish was a member of this "Class of '67," in on the ground floor of a list that now includes some 1,600 species of animals and plants.[12]

Udall developed the list after the enactment of the Endangered Species Preservation Act, passed by Congress and signed by President Lyndon Johnson in October 1966. A forerunner to the much more powerful 1973 Endangered Species Act, which continues to undergird endangered species policy, the 1966 law bluntly explained that "one of the unfortunate consequences of growth and development in the United States has been the extermination of some native species of fish and wildlife." The law then required that the secretary regularly consult with scientists regarding species whose "existence is endangered because its habitat is threatened with destruction, drastic modification, or severe curtailment, or because of overexploitation, disease, predation, or because of other factors."[13] The results of these consultations would form the basis for the listings.

The 1966 law, and its mandated list, represented a new legal and cultural class of animal: endangered ones. Previous wildlife laws had either dealt with animals in broad categories, such as migratory birds or game animals, or focused on a particular species, like a law passed in 1940 protecting bald eagles.[14] Here, for the first time, the federal government classified species on their shared future: endangered animals might not have one.

The act's preamble made it seem radical: industrial capitalism in the United States destroyed species and the federal government had a responsibility to stop that from happening. Period. The actual text of the act, though, was much more conservative. Focused only on animals (plants were not eligible), it directed federal agencies in the Departments of the Interior, Agriculture, and Defense—a list whose subagencies included the Park Service, BLM, and Forest Service—to "insofar as practicable . . . preserve the habitats of such threatened species on lands under their jurisdiction."[15] Basically, the law asked federal land management agencies to protect habitat, if "practicable," and only on lands they oversaw. If it became impractical, or a threat originated elsewhere, the species was out of luck.

Congress did not independently identify the need for endangered species legislation. Udall and the Department of the Interior had actually developed model legislation for Congress's consideration and had been working on a departmental list of endangered flora and fauna since 1964. That year,

Udall had directed the Bureau of Sport Fisheries and Wildlife to begin compiling a list of species threatened with extinction. Udall explicitly copied the model of a United Nations–inspired organization called the IUCN, or International Union for Conservation of Nature, that in 1961 started a color-coded system of plastic binders to track projects on conservation. In its "Red Book," it began collecting information on global species threatened with extinction, which included characteristics, distribution, estimated populations, reasons for decline, proposed protection measures, and references to scientific articles on the species. Returning from an IUCN meeting in Nairobi in late 1963, Udall brought the idea of developing a similar list for U.S. wildlife.[16]

The pupfish found its way onto the BSFW's growing list, it seems, thanks to a familiar name in the story of pupfish persistence: Robert Rush Miller, the former graduate student at the University of Michigan who in the 1930s and 1940s scoured the desert for fishes and put Ash Meadows on the ichthyological map. After a short stint at the Smithsonian, Miller joined the faculty of his alma mater in the late 1940s and from that position began serving as the chair of the IUCN's freshwater fish group in 1965. He was also tapped by BSFW to help develop the list of rare and endangered fishes after his outspoken criticism of the agency's Green River poisoning a few years earlier.[17] In January 1966—ten months before the passage of the country's first endangered species law—he provided the data that led to the Devils Hole pupfish's inclusion in the BSFW's Red Book. No specific threats to Devils Hole were identified in the entry. It simply stated, "Any species as restricted in distribution and number as is this one must be considered to be threatened."[18] When the Department of the Interior made its selections for the first class of endangered species, the Devils Hole pupfish had two clear advantages: it was already known to the BSFW, and it lived only on federal (NPS) land, where the act could have some power.

Despite its inclusion, however, what is most notable about the pupfish's presence on the endangered species list in 1967 was how easy it could be to ignore the law. Interior published the first list of endangered species in the *Federal Register* several months before the Bureau of Land Management printed the details of its proposed exchange with Cappaert in the same publication. According to the 1966 endangered species law, BLM had a responsibility "insofar as practicable" to protect listed species. But the Devils Hole pupfish's presence on the list and BLM's responsibility under it went unmentioned in the agency's 1967 and 1968 announcements in the *Register* to move forward with the Cappaert land exchange.[19] Even NPS missed the significance: Death Valley superintendent John Stratton's letter to BLM requesting the withdrawal of 560 acres from the exchange did not mention that the Devils Hole pupfish was on the endangered species list.[20]

When the *New York Times* ran an Associated Press article announcing the publication of the endangered species list, it quoted Udall as saying, "An informed public will act to help reduce the dangers threatening these rare animals."[21] In 1967, however, not even NPS was "informed" enough to mention the Devils Hole pupfish's presence on the list to BLM.

To the Moon

While the federal government's new endangered species law did not halt the BLM exchange with Cappaert's company or play an instrumental role in protecting the Devils Hole pupfish, it did begin to shift the focus of some agencies and emboldened employees to investigate the status of endangered species. The April 1969 trip to Ash Meadows represented such an investigation. The gathered employees from various field offices and university labs then agreed that they should plan a workshop to "explain who is doing what, where and when on endangered desert fish species."[22] They scheduled a conference for mid-November.

Hosted by Death Valley National Monument in Furnace Creek, the meeting was a watershed in the attention paid by the agencies to desert fishes. The conference included representatives from NPS, BLM, BSFW, the California and Nevada fish and game departments, the Nevada Division of Water Resources, as well as from seven different public universities in California, Nevada, and Arizona and a handful from private conservation groups—in total more than forty people.[23] (While this was an exciting development in the desert fishes program, the meeting could only place a distant second on a list of noteworthy events in Death Valley in the fall of 1969. A month earlier, in October, NPS rangers played a key role in the arrest of Charles Manson and his "family" on a ranch at the border of the monument. The crime that stoked the ire of the Park Service and led to the capture of Manson was the arson of a new articulating loader purchased by the Death Valley maintenance department.[24])

Over two days, the meeting explored the status of fish populations around the Death Valley region, existing and potential research projects, water rights and land management, and possible locations for transplanting threatened fishes. There was a lot to talk about: day one adjourned at 11 p.m.; day-two sessions started at 8 a.m. It also included a field trip to Ash Meadows, which revealed further changes by Cappaert's Spring Meadows Inc. in the six months since the April tour. They noted especially that the Ash Meadows Amargosa pupfish population at Jackrabbit Spring had been destroyed when a pump that was sunk directly into the spring dewatered the pool.[25]

Despite the gloomy circumstances for the meeting, what a far cry it was from 1950, when few at the Park Service backed the protection of Devils Hole. Carl Hubbs, who was in attendance, must have been proud to look at the sea change taking place at NPS and other agencies, as well as in university science departments in under twenty years.[26] From the fringes of fishery science and management, desert pupfish began speaking to a wider circle of scientists and managers, mirroring a growing environmental consciousness in the nation as a whole.

Perhaps no one from this younger generation better embodied or was more affected by the growing interest in desert fishes conservation than Edwin P. "Phil" Pister. A California Fish and Game fishery biologist based east of the Sierra Nevada Mountains in Bishop, the largest town in the Owens Valley, Pister was responsible for managing trout for sport fishing in nearby eastern Sierra Nevada lakes. This was what state fish and game agencies did, manage fish for people. The impulse that led BSFW to poison the Green River to aid the introduction of rainbow trout led Pister's agency to plant trout in high alpine lakes in the Sierra—even by dropping them from airplanes.[27] In the summer of 1964, however, Carl Hubbs and Robert Miller came to Bishop to see whether the Owens pupfish still lived in the valley, or if they had been made extinct by the diversion of water from the valley to Los Angeles. Pister's bosses let him tag along. In a marsh just north of Bishop, Hubbs saw the fish, yelling to Miller, "Bob, they're still here!"[28]

Years later, Pister described the experience of rediscovering the Owens pupfish as akin to being Paul on the road to Damascus.[29] He even told Miller that it was the most significant day of his career, and perhaps life.[30] "What I'd been doing up to that point was essentially devoting my life to providing freezer boxes full of trout for people to take home and cook in Los Angeles, and it wasn't doing a thing for the basic biological resource."[31] Pister continued his work on sport fishes but gradually became more and more invested in endemic, nongame fish issues. Famously—at least in fishery management circles—when the habitat near Bishop where Hubbs, Miller, and Pister had rediscovered the Owens pupfish began to dry up in the summer of 1969, he collected as many fish as he could and literally carried the entire known population of Owens pupfish in two buckets to another nearby pool.[32]

Months after his scare with the "species in a bucket," Pister brought tremendous energy to the November 1969 symposium, editing its proceedings, and helping to push it to become a permanent group called the Desert Fishes Council, a collaboration of university scientists and public agency leaders that still exists and for which he served as president and then as the long-time executive secretary. When I attended the council's fiftieth annual meeting in Death Valley in 2018—held in the same room as the first meeting—Pister was there, still participating, a few weeks shy of his ninetieth birthday. (When I began writing about the pupfish, Pister also welcomed me to his home in Bishop and shared his invaluable personal papers and color slides with me. These are used in several places throughout this book.)

While Pister is exceptional in his dedication to desert fishes, his attitude about the need to preserve them was something that appealed to a growing circle of agency staff and scientists in the late 1960s, tired of focusing narrowly on aquatic species people liked to catch, and eager to connect their experiences in the desert with broader environmental problems. In an introduction to the proceedings from the 1969 meeting, which he distributed in early 1970, Pister drew a link between the problems facing desert fishes to the "problems of environmental protection and preservation in the face of almost unbelievable social and technocratic problems." Just months after the Apollo 11 lunar landing mesmerized the world, he wrote, "If we are able to muster the necessary finances and technology to place an astronaut on the moon, we most certainly can devise a means of preserving in acceptable condition not only pupfish habitat but the remainder of the earth's environment as well."[33]

• Best-Laid Plans

Despite the bold words from Pister and the enthusiasm from the assembled crowd, there were many notable absences from the November 1969 symposium, absences that were nearly as important as who showed up. Neither the 1966 endangered species law nor a 1969 act that updated some of its provisions required very much of federal agencies.[34] The language "insofar as practicable" meant "optional" to many, and getting agencies to coordinate for the protection of pupfish would require some high-level support. Unfortunately, among federal officials, only one Interior Department employee made the trip from Washington, where all the agencies' leaders worked.[35]

Without backing from Washington—and thus little administrative support for attempting to protect the Ash Meadows habitats—the symposium concluded with an "action plan for species preservation" that focused primarily on identifying potential habitats for transplanting isolated fish species where they could be protected from extinction if their native habitats were destroyed.[36] Much like the 1966 Endangered Species Preservation Act, which had a critical framework but modest authority, the first symposium on desert fishes paired a hope that desert fish habitats could be saved with little to say about exactly how that would happen. The basic dilemma for Pister and others after the 1969 conference was overcoming this lack of enthusiasm, getting people who did not attend the meeting to pay attention to it.[37] The water rights system was not the only place where pupfish were invisible.

The first interest from the Interior Department in Washington came from Leslie Glasgow, assistant secretary in the department overseeing fish and wildlife, parks, and marine resources.[38] Glasgow was an Indiana native who had been tapped by the Nixon administration to leave his post at Louisiana State University for another stint in government service (he had previously led the Louisiana's state wildlife agency). Glasgow toured the springs and observed the agricultural development in the area in February 1970. Death Valley's chief naturalist, Peter Sanchez, who accompanied Glasgow, also observed new construction since the end of 1969, describing in a report that he saw expanded fields, new pumps installed on wells, and a half-mile-long earthen dam under construction near Point of Rocks.[39]

Glasgow returned to Washington alarmed at what he had witnessed. Writing to Commissioner of Wildlife Charles Meacham, he explained that Spring Meadows Inc.'s development might "result in the extinction of some species of desert pup fish [*sic*]," an outcome he considered a "disaster." He told Meacham to direct the Park Service and the Bureau of Sport Fisheries and Wildlife, along with cooperation from BLM, the state fish and game agencies, and university personnel, to produce a study and "make recommendations for saving these fish."[40] This was exactly what the participants at the November symposium needed.

Glasgow's memo came to rest on the desk of John Gottschalk, the director of BSFW. Gottschalk did not jump at Glasgow and Meacham's directive to assemble an interagency group to figure out how to protect the pupfish from extinction. In fact, his reply was a long-winded bureaucratic way of saying "no" to his bosses' request. "Interior agencies," Gottschalk explained to Glasgow, were already working on the issue to "the extent of their abilities and within the limitations of manpower and money."[41] He cited the BSFW's participation in the November symposium, its cooperation with state fish and game agencies, and its funding of some research on nongame fishes in the Southwest. And to provide further evidence of his agency's commitment to desert fishes, he attached a six-page report on the November meeting. As a demonstration of the agency's interest in protecting Ash Meadows, it was less than encouraging. The report had actually been written by Robert R. Miller under contract with NPS. Gottschalk's office simply retyped it (*sans* author and funding agency) and forwarded it along with its letter to Glasgow, passing the report off as a BSFW product. The situation in Ash Meadows, Gottschalk assured the assistant secretary, was "much brighter than when we commenced."[42]

Others inside BSFW were not so sure. In fact, Gottschalk's attitude made at least one of his employees nervous—as much for his own job as for the pupfish. J. P. Linduska, head of the bureau's nascent endangered species program, worried that Gottschalk was "not responsive" to Glasgow's request. He wrote to a colleague: "If one of these spp. [species] (the last individual) winds up parking on a sandbar, I don't want Glasgow saying 'why the hell wasn't I informed?'" While Linduska doubted that the agency would actually act on any plans delivered to Glasgow—he thought efforts to save the fish would end up being "big, involved and costly" and unlikely to be implemented—he did believe that "we'd better be on the record with a real package of recommendations."[43]

In early May 1970, nearly three months after Glasgow's initial memo requesting an Interior Department investigation, nothing had been done. And as disturbing as the development of Cappaert's ranch was for field employees to witness, the lack of response from Washington must have been excruciating. Robert Murphy, the Death Valley superintendent, wrote to the NPS director that he feared that the Devils Hole pupfish would be extinct by late 1970 and that immediate action was still required by the Interior Department to prevent this "catastrophe." "Pressure from the scientific community and interested citizens for action is mounting," Murphy wrote. "The big guns are trained on us."[44] Not on Cappaert, Murphy said, but on "us," the government. The persistence of the pupfish would turn not on what a private company did, but how public agencies responded.

Pressure from outside the agencies may have tipped the scales and forced action from Interior. During the 1950s and early 1960s, NPS and some scientists believed that keeping the pupfish out of the headlines was key to protecting it from human disturbance, but by 1970 they realized that the public could play a key role in protecting it—including from the indifference or caution of the agencies themselves.[45] Pister summarized the changed attitude after the November meeting: "Of extreme importance will be our ability to communicate the overall problem to the public who, in the final analysis, will decide whether agriculture, pupfish, or a compromise of both will prevail in the Death Valley system."[46] At a May 7 meeting in Washington called by Glasgow with representatives from BLM, BSFW, and NPS, attendees discussed a litany of publicity, much of it generated by participants in the November meeting: a special issue of the magazine *Cry California* focused on Ash Meadows, letters urging action from the public, a telegram from the Sierra Club, and the prospect of a forthcoming NBC documentary on endangered species, including the pupfish.[47] All of this, two NPS employees reported at the meeting, "made it evident that a Departmental position and action program must be formulated as soon as possible."[48] This time, Glasgow got what he asked for: a new department task force.

Placed at the head of the new group was James McBroom, an assistant director at BSFW. McBroom later recalled that when Commissioner of Wildlife Meacham tapped him for the position, Meacham said, "I needed the roughest, toughest man I could think of to head the program. It turned out to be you." McBroom considered Meacham's words to be of "doubtful flattery," but he did set about building a program for saving the Devils Hole pupfish and Ash Meadows.[49] At his disposal, McBroom had Washington staff from BSFW, BLM, NPS, USGS, and the Interior Department's office of the solicitor, in addition to the existing cooperation from the state fish and game departments of Nevada and California and already involved university faculty. The cavalry had arrived.

More than a year after the April 1969 field trip that revealed to scientists and managers how the logic of beneficial use would transform Ash Meadows, McBroom held the federal government's first real meeting to plan how to protect the fishes living there on May 25, 1970, at Death Valley monument headquarters in Furnace Creek. The point of the meeting, McBroom informed the assembled crowd, was twofold:

(1) to provide an interagency action plan by June 1, 1970 for the preservation of the pupfish.

(2) to obtain ideas and recommended action programs from all persons attending this meeting.[50]

In other words, after a year of field visits, memos, and alarm ringing by concerned scientists and employees, the federal government was finally ready to say the following about the pupfish: (1) we need a plan, preferably in seven days. And (2) does anyone have a plan, especially one that can be ready in seven days?

• Let's Save the Desert Pupfish

In reality, developing a coherent strategy for protecting the pupfish took much more than a few days to organize. As with the formation of the task force itself, the choices made to protect the Devils Hole pupfish in both the near and long term remained an improvised affair, even if the new group wanted a clear plan developed quickly. What was easy to see, document, and visualize in chart-form was the inevitable outcome at Devils Hole, should the task force do nothing. A graph of the water data collected by USGS at Devils Hole from the late 1960s and early 1970s detailed a trajectory that bears an easy resemblance to the stock market in 1929 or 2008. The minimum water level recorded in 1970 was close to a foot below its level at the start of the year—even the USGS water recorder that documented this decline had to be relocated to a deeper part of the pool after it was left "high and dry" in August.[51] It only got worse from there: the new recorder ultimately documented an all-time low in 1972 that was more than 2 feet below the pool level in the mid-1960s (fig. 4.2).

The declining water level in Devils Hole left behind a distinct bathtub ring along the walls of the cavern and quickly began exposing parts of the shallow shelf typically hidden below the waterline. The pupfish spawn exclusively on this shallow shelf of rock at the southern end, an area just 19 feet long and between 6.5 and 10 feet wide. At its minimum in 1972, less than 15 percent of the shelf remained covered by water, though removal of rocks on the shelf improved that situation somewhat.[52] If it continued to decline, the entire shallow shelf would be exposed, eliminating all space for pupfish

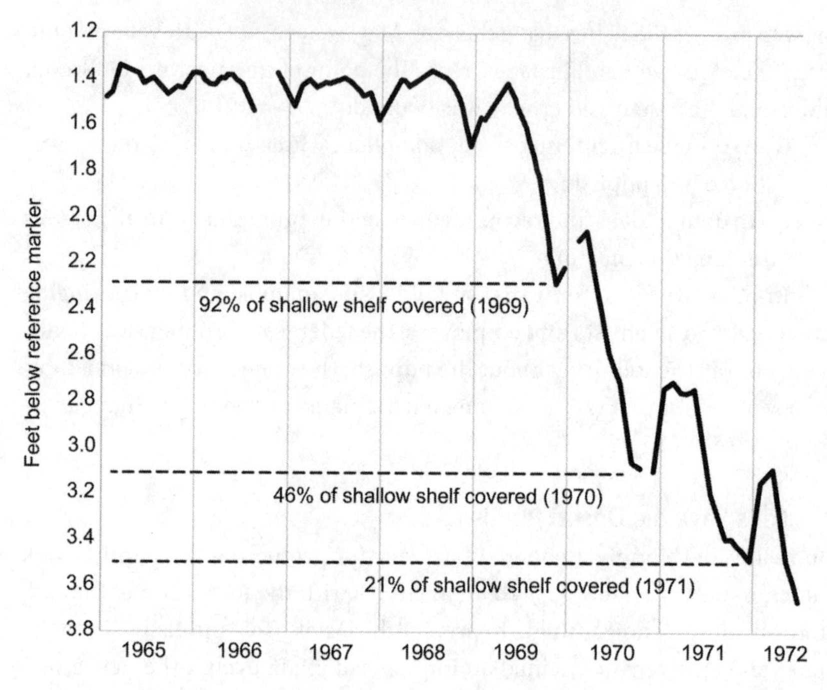

FIGURE 4.2. Monthly lowest water levels in Devils Hole, January 1965 through mid-1972. As the water level declined, less and less of the shallow shelf—on which the Devils Hole pupfish reproduce—remained covered by water. Chart from W. W. Dudley and J. D. Larson, *Effect of Irrigation Pumping on Desert Pupfish Habitats in Ash Meadows, Nye County, Nevada*, Geological Survey Professional Paper 927 (Washington, D.C.: GPO), 1976, p. 3.

reproduction, and in less than a year—the typical lifespan of a pupfish—the species would be extinct. (In a strange way, this is reminiscent of what Lake Mead, just east of Las Vegas, looked like as I started researching this book in southern Nevada in 2014—well below capacity, with another white bathtub line higher up on the canyon wall. One spelled disaster for the pupfish, the other, a still unfolding crisis for water users across the Southwest.)

The first achievement of the task force was an eight-page report titled, awkwardly, "A Task Force Report on Let's Save the Desert Pupfish," published in June 1970. Designed for distribution to the public—which continued to write to the federal government about what was being done to protect the fish—it explained that the Department of the Interior had grand plans for all the pupfishes of the Death Valley region, not just at Devils Hole. Drawn from its crash meeting in Death Valley at the end of May, the report mentioned asking the State of Nevada to limit groundwater pumping and support research on pupfish, as well as two items it would pursue "assuming the availability of funds": a detailed study of groundwater and acquiring privately owned land near some other pupfish habitats in Ash Meadows.[53]

A number of the participants in the task force meeting in Death Valley knew what they wanted—to stop pumping in Ash Meadows and thus halt the water-level decline in Devils Hole. But neither the meeting in Death Valley nor the report issued the following month explained clearly how that could happen. The task force, while it finally coordinated several agencies and scientists, still lacked any legal authority or funding to stop the water in Ash Meadows from being put to beneficial use. In a memo on the May meeting to his superiors at the Park Service, Death Valley's chief naturalist, Peter Sanchez, offered a much different take from the sunny presentation in the public-facing report. He wrote bluntly, "No recommendations for preserving the native habitat of Devils Hole were made."[54]

The phrase "assuming availability of funds" used in the report also stuck out to readers. Richard G. "Dick" Miller, the biologist who in 1948 had filed a water right application for Devils Hole in order to protect the pupfish, went on to found an environmental organization called the Foresta Institute for Ocean and Mountain Studies. Based in Nevada, his organization served as a watchdog and critic to federal environmental policy in the state during the 1960s and 1970s. One of Dick Miller's employees, Tina Nappe, attended both the November 1969 symposium and the May 1970 task force meeting. (At both events, she was one of the few women present.) Afterward, she reflected that the meeting was "notably lacking...citizen input," and with "agency people...confined by their positions," a discussion of how to secure a future for the Devils Hole habitat was "not broached at all."[55] Later, she wrote to McBroom that she felt "disappointed that nothing more conclusive was presented.... So long as the water declines, of what value are your efforts?" Prophetically, Nappe told McBroom she "realize[d] this places the Bureau in a very uncomfortable position with the only options appearing to be buying the ranch or bringing suit against the State of Nevada or the ranch."[56]

If Interior still edged away from such eventualities, it did begin to make a dent in its list of action items. It stopped making things worse by spiking further work on the Cappaert land exchange deal. In 1969, Cappaert had received the deed for the first 5,645 acres of the land swap, but an additional 7,307.99 acres he planned to acquire in Ash Meadows remained tied up in the Nevada BLM land office as employees sorted out whether valid mining claims still existed on them. This delay—which had nothing to do with the pupfish at first but rather the legacy of liberal nineteenth-century mining laws—ultimately worked in favor of pupfish conservation. In September 1970, the Interior Department ordered BLM to cease work on the exchange.[57] Additionally, after much handwringing about the cost and which agencies should bear it, the federal agencies funded and tasked USGS with an elaborate study of the groundwater resources in the Ash Meadows area, focused especially on how pumping from different wells would affect Devils Hole and the rest of Ash Meadows. This, in turn, might be the basis for legal action—a step that Interior's solicitor's office said required hard evidence that Cappaert's company was responsible for the water decline in Devils Hole.

Sorting all this out would take time, and many feared the Devils Hole pupfish would become extinct in 1970, since the task force still had no long-range plan for stemming groundwater withdrawals. At the second symposium on desert fishes in November 1970, Dick Miller's protégé Tina Nappe read a statement summarizing the situation: "No fish has before received the attention and investment of money and time that has consumed much of all our thoughts and efforts in the past year. . . . Yet pupfish-saving is essentially a failure."[58] In fact, nearly three years passed from the initial task force meeting to the first limits on pumping in Ash Meadows, and almost another five before a final settlement was reached.

Alongside the bigger question of long-term pumping cessation to protect water resources for the pupfish, keeping them alive in the short run also presented challenges. No one knew how to raise pupfish in aquaria or artificial refuges, nor did they even know much about how the pupfish used its natural habitat or how to improve it. Essentially, managers still faced the same question even after creation of the task force: how could they save the pupfish?

• That Says It All!!

Among the hundreds of slides Phil Pister took during the 1970s, I found one of his colleague James Deacon from the University of Nevada, Las Vegas. It is the image at the start of this chapter (fig. 4.1). Deacon leans against a rock at the foot of Devils Hole facing the camera. Evidence of conservation measures taken by NPS and the task force can be seen in the background, along with the partially exposed shallow shelf. A bank of lights hangs over the water, two large pieces of Styrofoam float in the deeper part of the pool, suspending an unseen wooden platform just below the surface. A Park Service employee checks the water level at the USGS recorder. Deacon's eyes are nearly closed, and his hand is on his forehead, his expression a mix of exhaustion and concern. On the cardboard edge of the 35 mm slide, Pister later wrote, "That says it all!!"

This photograph is remarkable in communicating some of the difficulties of "saving the pupfish"—it depicts the experimental mitigations that managers developed for the declining water level in Devils Hole and also conveys the seeming weariness of one of those who had devoted substantial amounts of time to the effort. Deacon, in particular, played a pivotal role in defining the minimum habitat requirements for the Devils Hole pupfish's survival, a development discussed in the next chapter.

A different, broader measure of the work that the managers and scientists affiliated with the Interior Department task force and the Desert Fishes Council undertook during this period is their volume of correspondence. Just *my notes* characterizing the memos and reports written about managing the pupfish during this period run over 75,000 words—nearly as long as this book. What all these memos detail is the collaborative invention of a crash program in endangered species conservation.

I say "crash program" because scientists and managers faced two serious limitations in developing interventions to forestall extinction. First, the total knowledge about Devils Hole accumulated since the Death Valley Expedition pocketed a handful of pupfish in 1891 was very particular. Historically, the question of interest to ichthyologists was about classifying desert fishes, charting their differences, and establishing a portrait of evolution in these extreme environments. This encouraged the method of "knowing" the pupfish through examining and measuring dead bodies, conveniently preserved in jars of alcohol on museum shelves. This process identified unique features of the pupfish in Devils Hole, resulted in its naming as a species, and provided a rationale to Hubbs and NPS for Devils Hole's protection in Death Valley National Monument.

This type of science made the pupfish worth saving in the 1960s and 1970s, but it did little to explain the interactions between the pupfish and its environment. Nor did it examine population dynamics, identify food sources, or establish methods for raising them in captivity. In other words, nearly eighty years after the first scientific visit to Devils Hole, no one really knew much about the pupfish as living, breathing animals or how they interacted with their desert home. At the November 1969 symposium, the lack of research was discussed openly: "Our knowledge of the taxonomy of the Death Valley system fishes greatly exceeds our knowledge of their biology," Pister summarized in the meeting's proceedings. And yet helping the fishes survive seemed to require actions—like maintaining habitats or building sanctuaries—that were "completely dependent upon the extent of our knowledge of the species involved."[59]

At the time that biologists and the Park Service began to worry about the pupfish, a new resource was hot off the presses: the first-ever study of Devils Hole pupfish ecology, a master's thesis written by a graduate student named Carol James under the direction of James Deacon at UNLV.[60] James's project, with fieldwork conducted during 1967 and 1968, resulted in some basic data about the Devils Hole habitat, as well as pupfish population size, structure, individual growth rates, mortality, longevity, and thermal tolerances. Learning such things about the population, however, did not come without casualties for the pupfish. In 1967, as James began her project, a number of veteran pupfish researchers including Carl Hubbs and Robert Miller gathered at Devils Hole to again attempt to count the entire population and measure the length of individuals by capturing them alive and temporarily storing them in submerged capture-traps. Over two days, the group removed at least 464 pupfish from Devils Hole. But they returned just 248. The remaining number—mostly small, young fish—were gone, apparently eaten by adults while held in artificial containers that did not have places for small fish to hide. At great cost, researchers learned that Devils Hole pupfish predate their young if given an opportunity.[61]

Beyond the little that was known about Devils Hole pupfish, specifically, and desert fishes more broadly, managers and scientists looking to protect the pupfish as the habitat shrank also faced a second problem: there was scant guidance available for developing the kind of successful intensive conservation projects that the task force needed. Conservation biology, the field dedicated to both understanding and managing endangered species and habitats, had not coalesced in 1970—the first workshop that began the process of defining the field took place in 1978.[62] While conservationists had sought to secure habitat since the early twentieth century to protect species, the best practices for intervening directly in degraded habitats, or how to conserve species through captive breeding programs, were not well developed, especially for nongame animals. The few existing intensive endangered species management programs formulated by the federal government dealt with birds such as the whooping crane.[63] (And what insights these projects might have had for other small populations had not yet been distilled.)

In this context, the scientific and management responses to the declining water level in Devils Hole were as improvised as the government structures in which they took place. Those who were eyeing the sagging water level in Devils Hole during the 1970s had little to go on. Neither scientists' history with the pupfish nor the established government or academic standards for intensive management of endangered species were particularly robust. The Devils Hole pupfish, then, survived because of the intense focus field employees at a number of public agencies brought to bear on the species during a transitional moment in American conservation policy. The pupfish may also have survived *despite* the specific actions taken on their behalf. Indeed, at the 2018 Desert Fishes Council meeting, Phil Pister reviewed the history of the interventions in Devils Hole during the 1970s before an audience of fishery biologists and managers who had for the most part only known the post–Endangered Species Act era, flashing slides of the work on the screen. In addition to hoping that the pupfish could survive the water decline, he told the audience, "We began to hope that the fish could survive the studies."[64]

The "studies" Pister referred to were two interventions visible in the background of the photograph he took of James Deacon in the early 1970s: the artificial shelf and lighting installed in Devils Hole. The basic idea, as Death Valley's Peter Sanchez explained in April 1970, would be for a submerged "floating shelf" to provide habitat for algae and aquatic invertebrates and space for pupfish to spawn as the water level decline reduced space on the natural shallow shelf.[65] Sanchez, along with UNLV's Deacon and Nevada Fish and Game's Dale Lockard installed the corrugated fiberglass platform just days after the announcement of the pupfish task force's creation.[66] Two months after the installation of the shelf, the task force decided to install a lighting fixture—a bank of sixteen 150-watt incandescent bulbs—over the pool to increase the amount of energy in the system and stimulate algae growth on the artificial shelf.[67]

While these interventions represented dramatic changes in the Devils Hole ecosystem, the task force and its cooperators made no systematic effort to monitor impact or efficacy. As early as December 1970, an NPS official in Washington expressed concern that "it has not been possible to adequately follow-up on the various management measures that have been undertaken during the past year." He hoped that more detailed monitoring of the population structure could be developed.[68] As if to validate his worry, between November 1970 and January 1971 estimates of the Devils Hole pupfish population by three different contributors to the task force resulted in such a wide spread—800–1,000, 225–250, and 116—as to render them useless even in guessing the impact of the shelf and lights, let alone determine the actual population.[69] Not until April 1972 did NPS begin organizing its employees under the supervision of Robert Miller to conduct monthly pupfish counts based on scuba diving—a practice that continues in similar fashion to this day.

With no regular monitoring of the habitat for almost two years after the introduction of the shelf and lights, problems could go unacknowledged. In 1970, NPS believed that a "superabundance of algae" noticed in Devils Hole was the "direct result of the artificial lighting."[70] By the spring of 1972, Robert Miller was having doubts that this was their main effect. In a letter to Phil Pister, he wondered "whether a continuous regime of artificial light is not having a bad effect on the species?"[71] When he visited Devils Hole to begin population surveys the following month, he found that the surface of the artificial shelf over which the lights were suspended was "covered with a thick, grey mass of dead algae," while the natural shelf continued to show signs of green algae.[72] Miller took a sample and brought it to an algae expert at the University of Utah who confirmed the problem: the Park Service had been using the wrong kind of light. Back in 1970, Miller had told the Park Service that they should use regular incandescent bulbs because they would be easier to install and maintain than "daylight" bulbs that only came in fluorescent tubes.[73] But, two years later, it seemed the use of the incandescent bulbs had changed the algae community in the pool. "We have been using the wrong light and too much of it!... How stupid can you get!" he wrote to Phil Pister. "We have been killing the algae with incandescent light."[74]

Miller was an expert on pupfish taxonomy, but neither he nor anyone else involved in the pupfish program knew much about algae. Like Miller, the university scientists were primarily ichthyologists, and the agency field employees were experts in fisheries or natural resources management. All it took to learn that "to culture algae one must use only fluorescent light" was a "series of telephone calls," Miller told NPS officials in Washington.[75] But those calls had not been made in 1970, a reflection of the intellectual backgrounds of those involved, the informal nature of the pupfish recovery program, and the urgency felt by the personnel involved to do *something*, even if it was not fully considered.

The pupfish task force was dissolved in the fall of 1971, but the programs it set in motion continued through a cooperative arrangement managed by the BSFW's Portland, Oregon office and through the continuity provided by the Desert Fishes Council. In the spring of 1973, as court actions (described in the next chapter) improved long-term prospects for the water level in Devils Hole, BSFW, NPS, and the Desert Fishes Council began reconsidering the infrastructure in the hole. "Outside of the esthetics of removing the shelf and lighting," a BSFW official wrote, "we feel it would be in the best interests, ecologically, to consider their removal." In contrast to their expected impact, he argued that the lights "did not significantly increase [algae] production" and that observations of the artificial shelf "suggest[ed] that its value was limited in the spawning and feeding of these fishes [*sic*]" and might even be blocking solar energy from diffusing deeper into the water column.[76] The "consensus" of the Desert Fishes Council's committee on Death Valley in 1974 agreed that the "lights have apparently served no useful purpose in Devils Hole" and that there was "no reason, then to keep them in place any longer." The value of the shelf was "questionable," and they had been "unable to correlate any population increases with it."[77] The lighting had already been turned off in 1973 (though the fixtures remained in place until the early 1980s), and the artificial shelf was removed in 1974.[78]

• Purgatory

All of this only partly explains the look of exasperation on James Deacon's face in the photograph at the start of this chapter. The agencies also undertook a second set of actions to keep the Devils Hole pupfish alive during the 1970s, one not as unfamiliar as the experimental light and shelf system: managers sought to raise the species somewhere else, either in aquaria or refuges (larger outdoor habitats). Ira La Rivers had moved Devils Hole pupfish to at least one other habitat in the 1940s in a clumsy effort to create a second population of the species. While Hubbs and Miller were skeptical of his actions at the time, Miller later proposed that because the Devils Hole pupfish is found in just one pool "it might be advisable, as a precaution, to attempt to establish this pupfish in another suitable habitat."[79] He repeated this proposal in a report to NPS in 1967.[80] NPS leadership concurred with Miller, but no action was taken to follow through on this idea prior to 1970.[81]

The first transplant of Devils Hole pupfish to a natural refuge since La Rivers's attempt(s) occurred with news cameras rolling in August 1970. Spearheaded by Pister and the California Department of Fish and Game, managers transplanted twenty-four pupfish to Upper Warm Spring in Saline Valley, a remote area wedged between Death Valley and the Owens Valley.[82] A few weeks later, the Nevada Fish and Game Department (NDFG) dug a ditch near Point of Rocks in Ash Meadows, piped in water from nearby springs, and deposited another two-dozen pupfish.[83] "We now have our 'eggs' in three baskets," NDFG biologist Dale Lockard reported to the task force, "which is heartening in light of the very marked water level decline in Devils Hole."[84]

In 1972, managers tried additional locations. The Park Service set off dynamite to improve a spring pool in the northern part of Death Valley near Scotty's Castle—an action that would be unlikely to move forward today as the land has since been designated a federal Wilderness Area. The Park Service later introduced twenty-five pupfish into the newly expanded pool.[85] BSFW took advantage of an unused artesian well drilled by George Swink in Ash Meadows. Biologists renamed the locale "Purgatory" to reflect the pupfish's plight and twice planted pupfish there.[86] Over both of these sites, the agencies built A-frame structures of lumber and canvas to regulate the light exposure and water temperature in an effort to make the refuges more like Devils Hole. In photographs, the structures look like strange, solitary tents left behind by prospectors from an earlier era in the region's history.

None of these four separate transplant sites proved successful in sustaining a Devils Hole pupfish population. (They did at least demonstrate the intensity of the collaborations between the agencies as engineered by the task force and the Desert Fishes Council: four different agencies had each taken the lead on one of the transplants.) In Saline Valley, in a farcical echo of the situation at Devils Hole, the water level in the refuge pool dropped after people rerouted the spring water away from the pool containing pupfish to feed a different pool they used for bathing.[87] In Ash Meadows, at the pit dug near Point of Rocks, the Ash Meadows Amargosa pupfish species managed to reach the pool, hybridizing with the Devils Hole pupfish stocked there.[88] Inside Death Valley National Monument at the dynamited pool, pupfish did not survive the high water temperatures or suspected predation by dragonflies.[89] Only at Purgatory, in Ash Meadows, did scientists observe some young pupfish—which indicated reproduction of a second generation. They believed, however, that Purgatory would "probably... not be self-sustaining," owing to predacious insects and algae choking the pool.[90] The population there did not survive.

As with transplants to refuges, the idea of raising pupfish in aquaria was hardly new. As an undergraduate in 1937, Miller wrote to Hubbs that he planned to establish a population of Devils Hole pupfish in an aquarium, a task he thought "should prove quite easy."[91] It was not; he failed. At least two other researchers also tried before 1970, but apparently neither succeeded.[92] By 1967, when Miller endorsed transplants to other natural habitats, he cautioned that the species "appears to be unsuitable for artificial [aquarium-based] propagation."[93] Under the threat from Cappaert's pumps, Miller and the task force felt they had no other choice but to try again.

The aquaria efforts fared no better than the refuges. Just before the creation of the task force, in the spring of 1970, NPS sent sixteen fish to an aquarium hobbyist who volunteered to attempt to spawn them. While the fish lived for a while, none of the spawning activity resulted in viable eggs.[94] In 1971, the agencies decided to go with professionals, sending fifteen pupfish to the Steinhart Aquarium, a part of the California Academy of Sciences in San Francisco. Steinhart's staff fared little better, able to rear just one pupfish.[95] And with the water level continuing to plummet in 1972, the Park Service sent another twenty-five pupfish to a California State University, Fresno, biologist. Of all, this was the most frustrating.

The fish arrived in the professor's lab in May 1972, but he was in the field for much of July and August. By the time he returned, twelve of the fish had died. By January 1973, all the fish were dead.[96] "Why didn't he notify us before we sent the pupfish to him, that he would not be able to tend to the fish during July and August?" asked a biologist at the Nevada Department of Fish and Game.[97] Instead of working on how to breed them, a miscommunication about the timing of the delivery of specimens resulted in another failure.

The transplant strategy finally paid off in the winter of 1972–1973, after scientists introduced pupfish to a concrete tank at the base of the Hoover

Dam, east of Las Vegas. The U.S. Bureau of Reclamation, which managed the dam, was (and is) a part of the Interior Department, and the agency was roped into participating in the task force, in part, because until the pupfish issue burst into Interior Department politics, the Bureau had been investigating the prospect of developing more agricultural land in the Amargosa Desert near Ash Meadows.[98] When the task force started thinking seriously about transplanting fish, the Bureau of Reclamation did a form of penance for its earlier plans by paying for the construction of an approximately 6-by-20-foot concrete tank, varying in depth from 3 feet on the shallow end to 9 feet.[99] The tank, perched on the Nevada side of the Colorado River, just downstream of the dam, was fed from a spring emerging from the canyon wall above. It was the first industrial recreation of the Devils Hole habitat—but not the last.

Twenty-seven pupfish were moved to the Hoover Dam refuge in October 1972.[100] Robert Miller, contracted by the Park Service to oversee the pupfish population counting at Devils Hole, visited the Hoover Dam site and in January 1973 said it "appear[ed] to be a successful transplant, with probably 100 or more fish" present in the concrete habitat.[101] By summer 1973, over 300 fish were reported to be in the refuge—more than were estimated to be in Devils Hole anytime during that year.[102] Whatever happened in Devils Hole, at least in the short term, the managers and biologists involved with the pupfish had succeeded in finding the pupfish another home.

The success at Hoover Dam also set a pattern in pupfish conservation that continues to this day, showing that establishing the species in a backup facility—keeping it in a kind of purgatory—was possible and prudent, though it has also been difficult to maintain. At Hoover Dam, the population of Devils Hole pupfish persisted uninterrupted until the mid-1980s when failures in the water supply system caused the population to crash.[103] Later restarted, the facility continued, on and off, to house Devils Hole pupfish into 2006. Similarly designed concrete tanks were built in Ash Meadows, one near School Spring, stocked with pupfish in 1980, and another near Point of Rocks, stocked in the early 1990s. (The fates of these refuges are discussed in chapter 6.)

Altogether, between May 1970 and September 1972, managers removed 196 pupfish from Devils Hole and attempted to breed them in aquaria and refuges.[104] The loss of virtually all these fish represented a stressful, slow-motion series of failures, undertaken out of desperation. And while the removal of pupfish for these transplant attempts never reached the single-day volumes of pupfish collected by Ira La Rivers, the total number of pupfish removed is significant—especially as during this period the habitat in Devils Hole was shrinking. Moreover, until April 1972 no effort had been made to accurately count the fish. But managers justified the decisions based on the continuing decline in the water level at Devils Hole. In early 1971, after the first attempts appeared to be failures, NDFG's Dale Lockard summed up the calculus: "We had little choice but to attempt to save this species in any way we could which included transplants into the best known possible sites available at that time.... This certainly involves risk; however, I still feel that we took what we felt to be the best courses of action at the time."[105] This situation allowed for the rapid implementation of efforts to "save the pupfish," but also fostered the implementation of ill-considered ideas such as distributing pupfish to an absent professor or assuming the interaction between artificial lights and algae was beneficial.

Today, managers would call all these efforts made to help the pupfish "recovery actions," a term of art from the post–Endangered Species Act of 1973 administrative world. That act created a process whereby all plans and actions on behalf of an endangered species are rigorously reviewed before implementation. In the early 1970s, however, pupfish conservation was the Wild West: the task force and its allies had little knowledge of Devils Hole ecology, no guidance from an academic field such as conservation biology, and the fish and habitat were both manipulated without the extensive formal review that characterize recovery efforts today.

The labor of field employees to make the federal government take actions on behalf of the pupfish—to recognize it—played a defining role in the species' survival of groundwater pumping by Cappaert's ranch. But their actions—both manipulating the Devils Hole habitat and transplanting fish to other places—did not on its own "save the pupfish." In fact, the results of these intensive maneuvers were primarily a single backup population at Hoover Dam. A second part of the federal government's commitment to pupfish came through in its legal challenge to Cappaert's pumping. But again, the drama was not just ranch versus federal government but the way in which federal authority interacted with the principles of Nevada's water law and its interpretation by state officials. The survival of the pupfish did not turn on Cappaert but on public agencies, the water rights system, and the relative power of the federal government and the State of Nevada.

CHAPTER 5

Kill the Pupfish

Dale Lockard served as a fisheries biologist at the Nevada Department of Fish and Game during the 1960s and 1970s. Like Phil Pister, his counterpart across the California state line, Lockard started his career in an agency with an interest in fish that seemed to extend only as far as species that weekend warriors wanted to find attached to their fishing poles. Beginning in 1969, however, Lockard spent substantial time focused on the very unfishable pupfish. He participated in the fortuitous April 1969 meeting in Ash Meadows where scientists and managers began to raise an alarm over the threat to the pupfish from Cappaert's Spring Meadows operation, and he later led pupfish transplant efforts, helped install the first artificial shelf in Devils Hole, and contributed to the Desert Fishes Council.[1]

Lockard did not act alone. The Nevada state legislature enabled his work with the pupfish and other endangered species. In spring 1969, Nevada became the first state in the nation to pass its own endangered species law, with the legislation's preamble closely echoing the federal government's 1966 statute in lamenting that "economic growth... has been attended with some serious and unfortunate consequences" for the state's native fish and wildlife.[2] Also like the first federal endangered species law, the Nevada statute was weak. It included exemptions to protect agricultural interests and ultimately played little direct role in the protection of the Devils Hole pupfish during the 1970s.

The statute's importance, then, lay both in giving cover for Lockard's work on the pupfish as well as in its broader symbolism: the act passed unanimously in the state assembly and by a wide margin in the state senate.[3] Writing in the pages of the *Las Vegas Sun*—in a column immediately adjacent to a report on the stocking of rainbow trout in Lake Mead—Lockard said that his department was committed to the "salvation" of endangered species and that it needed the "continued support of all Nevadans" to make it possible.[4] In principle, protecting the state's threatened endemic fauna was uncontroversial, if still incredibly difficult work.

Four years later, the situation had changed dramatically, and Dale Lockard became one of the first people working on pupfish preservation to realize it. In late July 1973, a letter arrived at the fish and game department and found its way to his desk. Included in the envelope was a bumper sticker. In large block letters, it read, "KILL THE PUPFISH."

Hollywood and detective fiction would have us believe that menacing notes arrive anonymously. But with the bumper sticker was a note on official letterhead from the Nevada legislature, signed by Tim Hafen, an assemblyman from Pahrump. In 1969, Hafen had been one of the legislators who unanimously passed the state's endangered species law. Just a few years later, however, Hafen was openly taunting one of the state agencies charged with implementing it. "Just thought I would send one of these stickers to give you an idea of local feeling. We feel the noise is all on the other side," he wrote.[5]

Stunned, Lockard wrote to Phil Pister. "How about this!" he exclaimed, enclosing photocopies of the bumper sticker and letter.[6] Lockard told Pister that he planned to leak the letter to the press, which he likely did since the *Sun* mentioned it a few weeks later in a story on a growing "anti-pupfish movement" in Nye County. "We don't really mean we should go out there and intentionally kill them," the sticker's creator and Nye County Commissioner Bob Ruud told the paper, "but when those people say 'Save the Pupfish,' they mean, by gosh, save it at any cost."[7]

At first glance, it is surprising that anyone, joking or not, would suggest killing the pupfish, since in contrast to mountain lions, rats, or coyotes—animals long persecuted in the West—pupfish do not bite or maim, serve as a disease vector, or prey on livestock. Right up until the appearance of the "kill the pupfish" bumper stickers, the central problem for those looking to "save the pupfish"—a phrase immortalized on its own bumper sticker in 1970—had been that people barely noticed them.[8] Indeed, much of the pupfish's twentieth-century history up to 1973 was a messy fight to keep it from being overlooked by scientists, citizens, the law, and government administrators. The pupfish persisted in Devils Hole both from—and occasionally despite—these efforts to define and recognize them as a species, as part of the national park system, and as a management priority across the Department of the Interior and state agencies.

An exception to this "recognition" story of pupfish survival, though, was water rights. Pupfish, as Richard G. Miller discovered in the late 1940s when he filed a state water permit application to protect the pupfish, were not a "beneficial use." Until George Swink, and then Francis Cappaert, imagined the potential of groundwater for agriculture in Ash Meadows this hardly mattered. The closest anyone got to putting Devils Hole water to beneficial use was John Bradford's aborted 1914 tunnel. The pupfish survived the first seven decades of the twentieth century on the outside of the water rights system. It survived being ignored.

But once Swink and Cappaert put in motion plans to develop Ash Meadows using groundwater, the Devils Hole water level would necessarily slump—something Cappaert's own water engineer suspected as early as 1967.[9] The fact that the water system did not recognize pupfish now threatened it with extinction, as legal water withdrawals for the beneficial use of raising crops and feeding cattle would ultimately "kill the pupfish," no bumper sticker necessary.

The "anti-pupfish movement" appeared only when the logic of beneficial use that excluded pupfish began to falter—when it appeared that electric pumps might *not* be able to finish the job. The month before Ruud's stickers began making the rounds, a judge had enjoined pumping from some of Cappaert's wells in order to protect the pupfish. In whatever way Hafen, Ruud, and others meant "kill the pupfish," the sentiment only made literal or rhetorical sense if the species was otherwise likely to survive. Judging by the attitudes of scientists and managers, this seemed tenuous from 1969 through 1972 (fig. 5.1).

This chapter, then, traces a long journey through the legal system as the pupfish became enmeshed in the administration of water rights in Nevada. It considers the bumpy start to the legal strategy of the pupfish task force and the unwillingness of the Nevada State Engineer to make space for the pupfish within the state government's water apportionment system. It then explores the federal government's winning legal argument and the proceedings that brought the case before the U.S. Supreme Court in 1976. Finally, it considers issues left unaddressed by the nation's highest court: just how much water would be needed in Devils Hole for the pupfish to survive and how would that affect future groundwater use around the wider region. Throughout these episodes, the survival of the pupfish hinged on an expansive vision of federal water rights and the ability of scientists to explain the interaction between groundwater pumping, Ash Meadows hydrology, and the Devils Hole ecosystem.

FIGURE 5.1. Bermuda grass at left, alfalfa at right on Cappaert Enterprises land in Ash Meadows, 1972. Digital scan by author of two photographs hand-stitched into a panorama by Bureau of Sport Fisheries and Wildlife staff, U.S. Fish and Wildlife Service office, Sacramento, CA.

"Kill the pupfish" bumper stickers were a direct reaction to this legal story, with resentment aimed at the limitations imposed by injunctions emanating from *federal* courtrooms. In rural Nevada, anger over the pupfish also foreshadowed a broader backlash against federal environmental and land policy near the end of the decade, the Sagebrush Rebellion. Put simply, the reason the pupfish survived the 1970s was also the reason it was wanted dead during that same decade. It could no longer be ignored by anyone, including those who wanted to use the desert's scarcest resource.

• The Litigious Pupfish

Lawsuits now play a central role in environmental protection in the United States. This was not always the case. As historian Paul Sabin has described, New Deal–era laws often made public agencies such as the National Park Service or U.S. Forest Service protectors of the public interest and the environment through their administrative actions—no lawsuits needed. Starting in the 1960s, however, a host of environmentally focused law groups, including the Natural Resources Defense Council and the Sierra Club Legal Defense Fund, began taking aim at these public agencies (in addition to private corporations) for failing to protect citizens and ecosystems from the ill effects of government planning and private profit-seeking alike.[10]

New environmental laws even enshrined a role for public-interest lawsuits ("citizen suits") in their statutes. Buried in the eleventh section of the Endangered Species Act of 1973 (ESA), for example, is a provision that allows any person or group to sue the federal government or private parties for violating the act.[11] Public-interest law provided both an impetus for such changes and, once enshrined in statutes, an approach that used the courts to challenge agency decisions on endangered species and the environment.

Some writers wrongly believe that the Devils Hole pupfish represents, in the words of one recent scientific article, "the first important test" of the ESA.[12] Yet the role of the ESA in protecting the pupfish during the 1970s was extremely limited. (And the first major court case revolving around the ESA focused on the snail darter, a species of fish that had not yet been discovered when the fight over the Devils Hole pupfish spilled into the courts.) As described in the previous chapter, though, in 1967 the pupfish was among the first seventy-eight species on the first list of endangered species announced in the *Federal Register*. The 1966 law that created the list was a forerunner to the ESA, but its modest authority did not extend to halting economic development on private property nor did it allow for citizen suits to enforce it. The much more robust ESA had both these powers, but President Richard Nixon did not sign that law until December 1973. For those looking for ways to ensure the pupfish's survival in 1969 and 1970, avenues other than endangered species law were needed.

Still, biologists involved in the nascent "save the pupfish" movement embodied the perspective of these new environmentally focused law groups. They proposed a lawsuit to halt groundwater pumping even before the federal government formed the task force in May 1970. In April of that year, veteran pupfish researcher Robert Miller wrote to a colleague that "nothing short of an injunction" would be necessary for the Devils Hole pupfish "to make it through the summer."[13] (The pupfish survived that year without a lawsuit.) The Sierra Club agreed on the need for legal action, even threatening to sue the Interior Department for its failure to, in turn, sue Cappaert to halt pumping.[14] Somewhere, it seemed, there must be a legal basis for preventing the drop in water level.

Lawyers in the Interior Department saw things differently from Miller and the Sierra Club. In fact, the department resisted starting a lawsuit for a simple but important reason: it did not know whom to sue. "An injunction could not be obtained," the solicitor's office explained in April 1970, "since there would be no specific person against whom to direct it."[15] USGS concurred, telling the task force that, while the timing of the water level decline and increase in pumping suggested that Cappaert's company was the "principal cause," it could not yet rule out other possibilities such as earthquakes or the numerous, ongoing nuclear tests at the Nevada Test Site, north of Devils Hole.[16] Even USGS's existing water level data collected at Devils Hole before 1962 was suspect, the agency admitted, since the water stage recorder had experienced various mechanical problems and, in what later became a recurring theme, had been vandalized at least twice.[17]

Just as NPS had been unable to respond effectively to the proposed land exchange between Cappaert and BLM because of how little it knew about Devils Hole in 1967 and 1968, the task force had a similar problem when it came to halting pumping. "A 'blanket' injunction against a group, based on the likelihood that one of its members *might* be responsible, would be unsustainable," the solicitor's office explained.[18] Lawsuits—winning ones, anyway—require evidence.

The task force first tried to satisfy the lawyers' objections on the cheap. In May 1970, shortly after its first meeting, the group paid $2,745 to a University of Nevada research institute for a new report on the effects of pumping on Devils Hole and Ash Meadows springs.[19] The task force set a deadline of less than three weeks for delivery, since the "Solicitor's Office... insisted on having this type of information to support any legal action."[20] The result was a thin, twenty-six-page report, cobbled together mostly from existing information.[21] Since the report did not actually contain any newly collected data to improve understandings of Ash Meadows hydrology and the effects of pumping, David Watts, an assistant solicitor, told the task force, it "does not provide a sufficient basis for the institution of litigation."[22]

If something cannot be done cheaply, sometimes a good backup option is doing it right. In the same months that the task force began installing lights and the artificial shelf in Devils Hole, and while they transplanted pupfish to springs around Nevada and California, they also raised $50,000 for USGS to undertake a detailed study of groundwater in Ash Meadows.[23] Leslie Glasgow, Assistant Secretary for Fish and Wildlife, Parks, and Marine Resources, explained that it "would be designed to establish with certainty" the relationship between groundwater pumping and the falling water level.[24] It was ultimately the single most important step taken by the task force to ensure pupfish survival. It was also a lot of work.

Starting in the fall of 1970, USGS detailed two employees to run the elaborate study. As mentioned in chapter 3, the Ash Meadows groundwater "flow system" comes principally from the north and east, after water collects as snow and rain in the Spring Mountains west of Las Vegas and then infiltrates soil and rock. In Ash Meadows, a fault pushed some of that water up into springs and formed Devils Hole. The trick for USGS was to understand how this general system worked and varied within Ash Meadows, and how specific wells being used by Cappaert's firm influenced the Devils Hole water level.

This proved to be no mean feat. William Dudley, from the USGS Denver office, designed the study. He had previous experience studying water in southern Nevada, having run the agency's groundwater monitoring program on the Nevada Test Site, where underground nuclear testing interacted with groundwater.[25] Jerry Larson from the USGS Las Vegas office assisted Dudley in the field. By winter, Larson was spending four days a week in Ash Meadows collecting data from twenty-one different water recorders installed across the landscape in order to document water levels in observation wells and the flow from springs. He also took manual readings of spring flow, and monitored electric power consumption and discharge from twelve Cappaert production wells.[26] Back in the office, Dudley started to analyze the meaning of all this data—"joyfully of course," he reported.[27] Dudley and Larson did not have preliminary data ready to share with the solicitor until the summer of 1971.

The slow and intense work needed even to accumulate the data on the area's hydrology created some doubt that a court case could proceed quickly enough to prevent extinction. "It is possible," Bureau of Sport Fisheries and Wildlife Region 1 director John D. Findlay wrote to the agency's Washington, D.C., office in early spring 1972, "that the only permanent solution" for the escalating ecological crisis in Ash Meadows "would be to acquire Spring Meadows... and withdraw the adjoining public land to preserve the several species of threatened desert fishes and their habitats."[28]

Indeed, during the early 1970s, two different plans to acquire Cappaert's land in Ash Meadows were proposed. Both floundered. A Senate bill from California's Alan Cranston languished because Nevada Senator Alan Bible, despite his conservationist reputation in establishing parks in other states, refused to even hold hearings on it before the parks and public lands subcommittee he chaired. Bible—whose law firm represented Cappaert's company until 1972—reassured a constituent who worried over limits on groundwater pumping, writing, "I have no reason to believe that any action will be had on [Cranston's] proposal."[29] Several iterations of a new BSFW wildlife refuge proposal, meanwhile, did not move forward because of their high cost and a lack of interest inside the agency for a wildlife refuge specifically for endemic and endangered fishes (the first such refuge would not be established by the agency until 1979 for the Moapa dace, north of Las Vegas).[30]

With possibilities for land acquisition dim, the conclusions of the USGS study were highly anticipated within the task force. Yet, even as managers waited to learn what USGS would be able to say about the effects of groundwater pumping on Devils Hole, a second question weighed on the pupfish task force. Even if the USGS analysis found Spring Meadows Inc. responsible for the water level decline in Devils Hole, was it illegal for them to do it? The first answer the task force received on the subject came at a water rights hearing before the state engineer and the Nevada Division of Water Resources in December 1970. It was not promising for the pupfish.

Nevada's system for administering water rights follows the "prior appropriation doctrine," and, as described, applying for a water right is a bureaucratic process with a set of clear administrative steps. In the spring of 1970, at the very time that federal employees forced their bosses to create the pupfish task force, Cappaert's Spring Meadows Inc. filed a bevy of additional water rights applications with the state engineer.[31] The applications—which either amended the planned location of pumps under existing applications, or requested new surface or groundwater resources—showed that the company was by no means done developing water in Ash Meadows, where it already had eight production wells and 3,000 acres under cultivation.[32] As part of the administrative process for new water permits, other water rights holders or interested parties could protest them before the Division of Water Resources, frequently on the grounds that approval of the pending application would impair an existing, older water right.

The National Park Service, of course, did not have an existing, state-recognized water right at Devils Hole, but it opposed ten of Spring Meadows Inc.'s permits anyway.[33] In August, NPS submitted its official protest. Later, task force chairman James McBroom made a specific request to the state engineer: institute a moratorium on new water applications in Ash Meadows while the USGS study continued.[34] Nevada's statutes, McBroom believed, provided that latitude.[35]

In response, the state engineer called a hearing on December 16, 1970, in Las Vegas. The hearing represented the first time since Richard G. Miller's 1948 attempt to secure a water right to protect the pupfish that Devils Hole had garnered so much attention from state water officials. Roland Westergard, the state engineer, had previously acknowledged that state law gave him the authority to delay or reject applications based on cases where "water supply studies" were being made or where court action was "pending," though such situations did not "necessarily require postponement."[36] At the hearing, an Interior Department field solicitor named Otto Aho proposed that Westergard give the task force until January 1, 1972 (just over twelve months away) to collect USGS's findings and "make presentation of the information for the use by the State Engineer" to decide on Spring Meadows Inc.'s new permits.[37]

The state engineer, however, did not want to wait. "You know what the decision has to be," he told the hearing room. "I am going to overrule the request of the government for an extension of time in making a determination." Westergard approved the new Spring Meadows Inc. permits, telling the Interior Department that it was "not in the public interest" to have a moratorium on new permits, considering that Spring Meadows Inc. intended to put that water to beneficial use and that there was "no record of a water right at Devils Hole" on which Cappaert's new permits would infringe.[38] As for endangered species law, which Aho asked about after Westergard announced his decision, Westergard explained he did not think it applied to maintaining particular water levels. The fact that the exchange between them never made clear whether they were talking about state or federal endangered species law hardly mattered; neither had the teeth to prevent further water withdrawals in Ash Meadows.[39]

By the end of 1970, the Devils Hole pupfish's status was the same as it ever was when it came to water rights: it could be ignored just as easily as it had been in the 1940s. "Kill the pupfish" bumper stickers would not be necessary, since the water rights system would take care of that soon enough. And aside from the pupfish's legal invisibility, the federal government did not yet have clear evidence to demonstrate that Cappaert's firm was responsible for shrinking the habitat. If Westergard's decision at the

December hearing was disappointing, Aho, the Interior Department lawyer, also thought it was "'totally expected' and was 'no surprise.'" The solicitor's office could have appealed Westergard's decision to the state court system. But Aho believed such judicial review would serve "no purpose" and that "it would be impossible to secure a reversal or modification."[40]

The Fort Belknap Inheritance

After its disappointing hearing before Nevada's state engineer, the pupfish task force did not give up on using a legal strategy to halt detrimental groundwater pumping in Ash Meadows and make a place for the pupfish within the water rights system. Instead, after USGS's preliminary results confirmed a relationship between Cappaert's pumping and the declining water level in Devils Hole, it pursued a case in federal courtrooms. To do this, Interior and Justice Department lawyers drew on an old legal concept they thought was readily applicable to the pupfish case: federal reserved water rights.

The concept emerged from a conflict almost a thousand miles from Devils Hole. In 1888, nearly three years before federal scientists first visited Devils Hole as part of the Death Valley Expedition, the federal government signed a treaty with two Native American groups, the Gros Ventre and the Assiniboine, creating the Fort Belknap Indian Reservation in Montana. The agreement, which created a reservation covering 600,000 acres south of the Milk River, represented just the latest in a series of encroachments and revisions that stripped territory from Native Americans living in what became Montana during the nineteenth century, and it represented another milestone in the broader dispossession, coercion, and violence toward Indigenous people that undergirds modern American history.

Fort Belknap may be far from Devils Hole, but the Devils Hole pupfish owes its survival, in part, to this place and a fight over water along the Milk River. In 1905, from a combination of drought and upstream water diversions by Anglo settlers, very little water reached the reservation's crops. The federal government, which oversaw the reservation through the Bureau of Indian Affairs, filed suit against the upstream diverters. By 1908, the case, *Winters v. U.S.*, had risen all the way to the U.S. Supreme Court. There, the justices found that when the 1888 treaty was signed by the Gros Ventre, the Assiniboine, and the federal government—even though water rights were not specifically described—the agreement had *implied* that water was "reserved" by Native Americans and the federal government in sufficient quantities necessary to accomplish the purposes for which the reservation had been set aside, namely agriculture. The actual amount of water reserved—and here lay the case's potentially radical implications—was not an amount fixed in 1888 but the quantity that might become necessary in the future to make the reservation (and its people) successful. And because of the first-come first-served nature of the prior appropriation doctrine (which Montana also followed), the reservation held a water right senior to all those upstream settlers who arrived after Fort Belknap had been created.[41]

The reserved water rights affirmed in *Winters* gave an expansive vision of both Native American and federal authority over water rights—a system typically administered for the benefit of Euro-American settlers by state governments. It did not matter that the reservation lacked an established, senior water right under state law. The agreement between the federal government (the sovereign) and Native Americans superseded state government administration.

The federal reserved water rights doctrine established through *Winters* was well-known among lawyers in the Interior Department by the time they were brought into the pupfish task force. Until the 1960s, though, this reserved rights doctrine had been applied exclusively to Native American reservations. But in 1963, another water rights case—*Arizona v. California*—reached the U.S. Supreme Court. As part of a complex decision that capped years of litigation over how these two states should share water from the lower Colorado River, the Court broadened the principle of reserved water rights, applying it to Lake Mead National Recreation Area, as well as a national forest and two national wildlife refuges. As with the signing of a treaty, the Court said that when the federal government set aside land for a particular reason, it also reserved the water necessary to accomplish that purpose.[42]

In July 1971, some seven months after failing to persuade the Nevada state engineer to institute a short moratorium on new water rights in Ash Meadows as USGS investigated the effect of pumping on Devils Hole, the Interior Department solicitor's office sent an eight-page letter to Attorney General John Mitchell and his Justice Department, requesting litigation against Cappaert's Spring Meadows Inc. to halt groundwater pumping and to assert federal reserved water rights, drawing on the logic of *Winters* and *Arizona v. California*. The letter articulated the major arguments that the federal government would use as it pursued the case over the next seven years. Essentially, the solicitor argued that following the "reservation doctrine," when President Truman added Devils Hole to Death Valley National Monument in 1952 specifically to protect the pupfish, the federal government had reserved enough water to accomplish this purpose. Cappaert's water rights were not filed until the 1960s, and though recognized by Nevada's state engineer, they were nonetheless junior to what the solicitor believed was the federal government's right at Devils Hole. Under the appropriation doctrine, the most senior water rights holders have power to exercise their rights before later appropriators. The establishment of a federal reserved right in this case would make Devils Hole one of the oldest water rights in Ash Meadows and the wider Amargosa Desert groundwater basin. And it would mean that the federal government had priority to use—or in this case, not use—enough water to protect the pupfish before groundwater could be pumped by Cappaert or other more recent appropriators.

Little about the case was legally novel, the solicitor believed. It relied on well-established water case law, not recently passed environmental laws. The solicitor mentioned that the Devils Hole pupfish was on the endangered species list, but this did not provide an argument for the courtroom, just a rationale for why the Justice Department should make this case a priority. About the only wrinkle, the solicitor's office said, was that previous reserved water rights cases had all dealt with surface water, meaning that the only new issue seemed to be the "extension of the 'reservation doctrine' to ground water resources." As the 1971 irrigation season had already begun, the solicitor noted, "your expeditious handling of the matter would be greatly appreciated."[43]

The Justice Department brought its case to the United States District Court for the District of Nevada a month later, in August 1971. The suit against Spring Meadows Inc.—newly renamed Cappaert Enterprises—requested that the judge, Roger Foley, issue an injunction "in order to protect the federal water rights in Devils Hole" by halting pumping from three of the company's wells that USGS's research indicated had a strong effect on the Devils Hole water level.[44]

Despite the immediate need for a remedy to protect the Devils Hole habitat—the water level continued its downward trend with only modest recoveries in the winter when pumping ceased—it was almost a year later, in early July 1972, that the Justice Department actually brought its entire case against Cappaert before the court. At first, extensive legal action was delayed because on the eve of the case's first hearing in September 1971, the Justice Department and Cappaert's attorneys reached an agreement called a "stipulation of continuance." In exchange for avoiding a court fight over an injunction, the company agreed to halt pumping on the three wells described in the Department of Justice's complaint. (It also agreed not to increase pumping on five other operating wells in order to make up for the loss or drill new wells in the same vicinity of the prohibited wells.)[45]

The Department of the Interior considered the agreement to be a moment of matching its "words with action" on the pupfish issue, according to a department press release.[46] Over the following months, however, USGS data showed that the benefits of shutting down the three wells had been extremely modest.[47] Additionally, evidence trickled into Interior suggesting that Cappaert Enterprises viewed the stipulation as a shell game: it drilled two new wells, upgraded the pumps on an existing well, and filed new water permits with the state associated with one of the prohibited wells.[48] These actions, a lawyer from the Interior Department solicitor's office commented, "did appear to be adverse to and in bad faith to the present stipulation of continuance."[49]

After word of the company's behavior spread—and USGS predicted that the water level in Devils Hole would fall low enough to expose nearly the entire shallow shelf in 1972—NPS and the Bureau of Sport Fisheries and Wildlife pushed the solicitor's office and the Department of Justice to reopen the case.[50] If the federal government went that far, one Justice Department lawyer argued, "we might as well move our entire case."[51] District Court hearing dates were set for July 3 and 5, 1972. "We have been advised informally by the various staff members involved with the case," an assistant solicitor commented dryly, "that the fish will still be alive by that date."[52]

The pupfish not only survived through 1972 but also made it to the spring of 1973 before Judge Foley ruled. It was worth the wait. His decision represented the first moment that the pupfish's existence had been recognized in the water rights system. Foley accepted the government's argument that when President Truman designated Devils Hole as part of Death Valley National Monument the "unappropriated waters in, on, under and appurtenant to Devils Hole were withdrawn from private appropriation as against the United States and reserved to the extent necessary for the requirements and purposes of the said reservation."[53] Furthermore, while endangered species law did not have any specific legal force to prohibit pumping, the attention of legislators and the Interior Department toward endangered species showed that protecting the pupfish was in the "public interest." This was exactly the opposite construction of the term that State Engineer Westergard had reached in his hearing in December 1970, where he explained it was not in the public interest to delay Cappaert's permits that would put water to "beneficial use."[54]

Foley did not find the potpourri of arguments made by Cappaert's lawyers very convincing. They had argued, among other things, that (1) when Devils Hole was added to Death Valley in 1952, it was done to protect the geologic formation, not the pupfish, so no particular water level should be maintained; (2) the reservation doctrine did not apply to groundwater; and (3) the federal government, since it had deeded some of the Ash Meadows land to Cappaert, should not be allowed to prevent his company from using the property.[55] (Cappaert's lawyers also later made claims that were awkward for the State of Nevada, which had joined the case on Cappaert's side after State Engineer Westergard told the state attorney general and governor that the case had a critical bearing on the "State authority over public resources."[56] Despite this new ally, Cappaert's attorneys argued that landowners owned the water beneath their property—a view that would undermine decades of Nevada law that declared water to be owned by the public and administered by the state.[57])

Judge Foley did more than deal with principles at stake. In addition to accepting the concept of federal reserved water rights at Devils Hole, he instituted an injunction against Cappaert's pumping. By June 1973, when the injunction took effect, Foley's order prohibited pumping from eight Cappaert wells "and all other wells now existing or hereafter drilled on designated sections of land...which would be detrimental to the water rights of the United States and the survival of the Devils Hole pupfish."[58] The preliminary injunction did not halt all pumping by the ranch, only that which harmed the pupfish. Foley set a minimum mean daily water level for Devils Hole that was enough to cover much of the shallow shelf that the pupfish needed for breeding, and he appointed a "special master" to keep a record of the water level as well as instruct the ranch on how much water it could continue to pump while keeping above this minimum.[59] Ultimately, Judge

Foley's decision meant that the pupfish for the first time had been recognized and incorporated into the water rights system. This development owed much to the efforts of scientists and managers to force agencies to pay attention to the species but also to histories much further from Devils Hole, especially Fort Belknap and the *Winters* decision.

People took notice. The "kill the pupfish" bumper stickers appeared shortly after Foley's decision. Columnist Jack Anderson opined in the *Washington Post* after Foley's preliminary injunction that when "forced to choose between 200 ugly, inedible pupfish or a millionaire GOP contributor, the Nixon administration has put the welfare of the pupfish first."[60] George Swink, the former owner of the lands in Ash Meadows that formed the basis for Cappaert's ranch, meanwhile, did not think the story was over for the "millionaire GOP contributor." "I haven't heard from Cap or any of his people for a long time," Swink wrote to State Engineer Westergard, "but they surely appeals that Las Vegas decision [*sic*]."[61]

• Long Live the Pupfish

Francis Cappaert proved George Swink correct—more than once. Cappaert not only appealed Judge Foley's preliminary injunction to the U.S. Court of Appeals for the Ninth Circuit—where he also received an adverse ruling—but he then sought a hearing before the Supreme Court.[62] In a unanimous decision, the Court ruled against Cappaert as well, and upheld Foley's District Court ruling. Most stories of the pupfish's conservation and survival during the 1970s end with this trip to the high court in 1976. But appreciating how the pupfish survived—and continues to survive—requires investigating a question on which the Supreme Court offered little guidance: How much water do pupfish need in order to endure in Devils Hole?

For scientists and fishery biologists who had spent years trying to protect the pupfish from groundwater pumping, the Supreme Court decision was, of course, a huge relief. In early June 1976, Carl Hubbs and Robert Miller were at the University of Alaska in Fairbanks attending the annual meeting of the American Society for Ichthyology and Herpetology, the organization Hubbs once headed.[63] While Hubbs and Miller participated in the week-long meeting, the Supreme Court announced its decision in *Cappaert v. United States*.

Concluding his opinion for the unanimous court, Chief Justice Warren Burger summarized the decision: "As of 1952 when the United States reserved Devils Hole, it acquired by reservation water rights in unappropriated appurtenant water sufficient to maintain the level of the pool to preserve its scientific value and thereby implement Proclamation No. 2961."[64] The federal government, by virtue of setting aside Devils Hole as part of the national park system through President Truman's proclamation in 1952, had a right to water sufficient to support the pupfish's existence. Cappaert and the State of Nevada offered objections over the inapplicability of the reservation doctrine to groundwater resources, as well as estoppel and jurisdictional issues, among others, all of which the court found unconvincing.

Phil Pister heard the news of the Supreme Court's decision from University of Nevada, Las Vegas, Professor James Deacon, who called Pister at his California Department of Fish and Game office in Bishop. Pister shut himself in his office and wept tears of relief.[65] Knowing Hubbs and Miller were even further from Washington than he was on the east side of the Sierra Nevada Mountains, Pister placed a call to the University of Alaska, ultimately convincing someone at the other end of the line to take down a short message: "Supreme Court decided 9–0 in favor of pupfish." After receiving the note, Hubbs interrupted the meeting's proceedings and asked to make an announcement. He read Pister's message to the room of fishery scientists and received a hearty round of applause. Hubbs added, "Long live the pupfish."[66]

It had been over seven years since Pister and other managers and scientists first assembled in Ash Meadows in April 1969 and observed the growing ranching operation and the well and pump near Devils Hole. In that time, the group had convinced the leaders of its agencies to pay attention to this situation—no easy task—and then transplanted pupfish, took drastic management actions in Devils Hole itself, funded a groundwater study, and watched as a legal case bounced around the federal court system. The survival of the Devils Hole pupfish remained questionable throughout this period. Now, finally, the U.S. Supreme Court affirmed the district and appellate court rulings. More than that, the Court affirmed that the Devils Hole pupfish had a place in the apportionment of water in the State of Nevada. And while they never quite said it, the reservation doctrine and the words of Chief Justice Burger suggested that the survival of the Devils Hole pupfish was indeed a "beneficial use" of Nevada's water.

The decision, however, was incomplete. Burger wrote, "The implied-reservation-of-water-rights doctrine... reserves only that amount of water necessary to fulfill the purpose of the reservation, no more." But Burger and the Court did not provide any clues on the amount of water that might be "necessary" at Devils Hole.[67] In other words, the court decided everything except what was perhaps the most important part of the case: how *much* water the Devils Hole pupfish needed to survive. It was up to scientists and Justice Department lawyers to make a case in Judge Roger Foley's Las Vegas courtroom as to the minimum water level for the pupfish to survive. Figuring that out started with an inexpensive hardware store item: a copper washer.

Installed in 1962, USGS pegged the washer into the wall above the pool in Devils Hole to serve as a reference marker and to calibrate the other water monitoring equipment.[68] All subsequent discussions of water level in Devils Hole—i.e., how much water the pupfish should be afforded within the water rights system—turned on measurements in relation to this little washer. The minimum water level in Devils Hole prior to groundwater pumping hovered between 1.3 and 1.5 feet below the washer. At its nadir in 1972, the water level fell to 3.8 feet. A 30-inch drop in water level from the pre-pumping period pushed the pupfish to the edge of extinction, as nearly the entire shallow breeding shelf was exposed (fig. 5.2).

As difficult as it can be to deal with the principles in water law—federal reserved rights, the doctrine of prior appropriation—applying them to the peculiarities of pupfish ecology also proved challenging. From a human perspective, the difference between the pre-pumping and 1972 water level in Devils Hole was the difference between wading in thigh deep water or standing in a puddle. For the pupfish—as well as water users—this difference was everything. The first preliminary injunction issued by Judge Foley in 1973 restricted pumping by Cappaert to the extent that the water level would not drop past 3.0 feet from the copper washer. Later that summer, the Ninth Circuit Court of Appeals in San Francisco allowed the minimum water level to fall to 3.3 feet, arguing this level better reflected the status quo while a final determination in the case was made.

The federal government began preparing to revisit the water level question even before the Supreme Court ruled. As early as February 1976, with the announcement of a decision still months away, Death Valley National Monument employees began discussing the need "to get, *now*, our information lined up and ready" on what the minimum water level in Devils Hole should be, since it was possible that the Supreme Court would rule "only

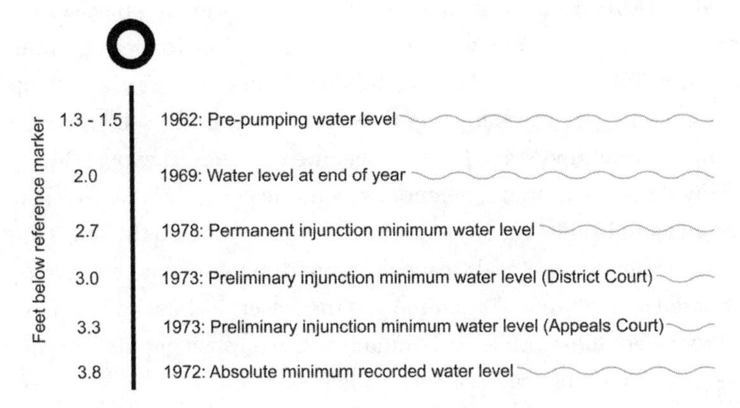

FIGURE 5.2. The measurement used in the *Cappaert* legal proceedings to calculate the extent of the federal government's right to water at Devils Hole was not computed in gallons, cubic feet per second, or acre-feet per year—the typical measures in water administration. Instead, the court's decisions focused on the water level in the Devils Hole pool itself, measured in tenths of a foot below a copper washer installed in the pool as a reference marker.

on the narrow issue of our right to the water," not any particular level.[69] After this prediction proved correct, NPS and Department of Justice staff believed that they were in a position for Judge Foley to set a permanent level of 3.0 feet below the copper washer, but not any higher unless additional evidence was forthcoming.[70]

Into early November 1976, then, it seemed that the federal government would continue the call for the 3.0-foot water level that Foley had decided back in June 1973 with his preliminary injunction.[71] On November 30, however, the Justice Department introduced a different motion in district court. Based on the "deposition of Dr. James E. Deacon, taken during the period November 23–26, 1976," along with other files and records, Justice filed a motion to modify the permanent injunction "so as to raise the water level permanently to 2.7 feet, the level recent studies have shown is necessary to insure the survival of *Cyprinodon diabolis*."[72]

James Deacon had first visited Devils Hole long before the November 1976 hearing. In February 1961, at the very start of what would be more than forty years as a faculty member at the University of Nevada, Las Vegas, Deacon joined a group of cave divers permitted by the Park Service to explore the site. In brief field notes from the trip, he estimated the pupfish population at more than 300 individuals.[73] When I interviewed him about his first visit to Devils Hole, he also admitted a more humbling piece of information. At the time, Deacon was a novice diver wearing loaned equipment. Underwater in Devils Hole, he became disoriented and nauseated; upon surfacing he became, perhaps, the first person to vomit in Devils Hole.[74] From this unassuming start, Deacon became not only an expert diver but also an avid investigator and defender of aquatic ecosystems across southern Nevada and Utah.[75] His student Carol James conducted the first ecological study of Devils Hole, and Deacon attended the fateful meeting in April 1969 that led to the pupfish task force and the Desert Fishes Council. In 1974, he had replaced Robert Miller in conducting the pupfish population counts for NPS. (Deacon is also pictured at Devils Hole in fig. 4.1.)

The federal government's decision to request a higher permanent water level in Devils Hole at a December 1976 hearing stemmed directly from Deacon's expertise and his observations there, which included population estimates and data he collected on light duration and intensity and changes in the water's dissolved oxygen concentration. This data made it possible for Deacon to determine "the primary productivity index of the area circumscribed by the natural shelf," as the government explained in its legal memorandum supporting its new claim.[76] Essentially, this "primary productivity index" measure calculated food availability for pupfish at different water levels. Along with the continued population monitoring and a detailed computer model of the shallow shelf, the work allowed Deacon to show a relationship between water level and population size.

Based on Deacon's study, the Department of Justice argued that "a minimum water level of 2.7 [feet] is necessary to insure [*sic*] a population of pupfish that would fluctuate between 200 to 400 individuals, the minimum threshold number of individuals necessary to insure the survivability of the species at Devils Hole." Cappaert's lawyers, meanwhile, proposed that 3.3 feet was enough. The species had not become extinct in the years since the firm's pumping began, so why should higher water levels be required? Absent any clear biological data of its own to present, and with little other outside evidence supporting its argument, Cappaert's case rested on casting doubt on Deacon's work. "The December 1976 hearing was substantially a one-witness hearing," Cappaert's lawyers concluded.[77]

How much habitat the pupfish needed to survive was the kind of specific question that a legal doctrine could not answer. Resolving the *Cappaert* litigation, then, ultimately depended as much on ecological monitoring and research, like the kind Deacon conducted for NPS, as it did on the principles of federal reserved water rights stretching back to the *Winters* decision. "After having read and reread the transcripts, studied and restudied the exhibits, and read and reread the briefs," Judge Foley wrote in December 1977, "this Court has concluded that plaintiff has established by a preponderance of evidence that its motion to modify the minimum water level at Devils Hole to 2.7 feet below the copper washer should be granted."[78] Long invisible, the pupfish now had a place in how the West's scarcest resource was apportioned.

• Rotenone

Because the implications of the Devils Hole pupfish's recognition within the water rights system had material consequences for water users in Nevada, the years of court hearings surrounding the *Cappaert* litigation attracted considerable attention. Reporting the results of the 1976 Supreme Court decision, the *Nevada West and Pahrump Valley Times* ran the headline, "17 States Lose Water Decision," referring to Nevada and sixteen other western states that filed briefs before the court on Cappaert's side.[79] Few people in the West, of course, had ever read the scientific papers by Joseph Wales and Robert Miller that defined the pupfish as a species, or knew of President Truman's proclamation that confirmed it as an object of conservation. But once those understandings of the pupfish as distinctive and worthy of protection merged with limitations on water use, the pupfish became a household name—and, to some, a threat.

The outcome of *Cappaert* had immediate and practical consequences. Supporting the water level in Devils Hole meant limiting pumping, both at the Cappaert ranch and in the surrounding area. The *Nevada West and Pahrump Valley Times* article detailing the outcome of the Supreme Court case quoted Cappaert's ranch manager, B. L. "Barney" Barnett as saying that before the injunctions began the ranch was still growing—employing 80 people, cultivating some 2,800 acres, and stocking 1,500 head of cattle. After the implementation of the injunctions that number shrank, according to Barnett, to 15 workers on 1,500 acres. The Supreme Court upholding the injunction was, Barnett said, "The rottenest damn deal I've ever run into in my life."[80] With the revised, higher, 2.7-foot minimum water level ordered by Judge Foley in 1977, even less pumping on the ranch would be allowed.

Anxiety about the pupfish case, however, stretched far beyond the fate of one ranch, evidenced by the anger in Pahrump that inspired Bob Ruud to print "kill the pupfish" bumper stickers and state assemblyman Tim Hafen to taunt Dale Lockard and the Nevada Department of Fish and Game with one in 1973. The federal water right at Devils Hole would limit the state's ability to manage its water for future economic development in the area surrounding Ash Meadows. Making water available for the survival of the pupfish meant less water available for visions of growth in southern Nevada. For farmers, developers, boosters, and administrators of Nevada's water system—those who had been dreaming of how to put the state's waters to "beneficial use" since the nineteenth century—this was a serious ideological sin.

As early as November 1972, months before Judge Foley issued his first injunction, a groundwater engineer at the Nevada Division of Water Resources, William Newman, attended the Desert Fishes Council meeting and captured this possibility in a letter to his boss, State Engineer Roland Westergard. "After being exposed to the discussions for the two days of this symposium, it is evident that, with all of the organizations involved in the study and protection of the environment, it will be very difficult to develop land to its full potential," he wrote. Given the extent of the public interest, ongoing litigation, and the hydrological and land acquisition studies underway, this was a fair analysis.

But Newman was not finished. "It should be with some concern that one contemplates the amount of federal and states [*sic*] funds that have been spent...on this rather minuscule program that, with all of its ramifications, has the potential to stymie development and...threatens the removal of existing development from the tax rolls and affect the economy of the State," he opined. "It would appear that the generation of power, a potential water supply for metropolitan Las Vegas or a food and fodder operation must also be important to the taxpayers' survival."[81]

By 1976, this feeling that taxpayers, developers, urban residents, and farmers were the endangered ones—not the pupfish—reached its rhetorical height in two key moments. First, before the Supreme Court, Cappaert's lawyer Samuel Lionel was the only person at the hearing to mention endangered species—but the potential applicability of the Endangered Species Act was not on his mind. He told the justices, "This case presents a contest between two endangered species. On the one hand, the individual farm operator of the United States whose numbers decline[d] no less than 40% between the years 1959 and 1969 and on the other hand a species of fish known as *Cyprinodon diabolis*."[82] While it was profoundly disingenuous to paint Cappaert, a multimillionaire industrialist, "as an individual farm operator," Lionel still captured the way in which the survival of the pupfish through its incorporation into the water management of Nevada did threaten long-held prerogatives to use resources for agriculture.

The second and more dramatic framing of Nevadans as the ones endangered by the pupfish came in the form of a newspaper editorial in the late winter of 1976, as the Supreme Court weighed its decision. In a March 8 editorial, Mel Steninger, publisher of the *Elko Free Press*, reviewed a recent speech by Roland Westergard, the state engineer, who had taken to calling the Devils Hole story the "Pupfish Caper" in reference to the supposed theft of Nevada's water by the federal government. In the future, Westergard cautioned, the "issue might be known as the Pupfish calamity or catastrophe."[83] Steninger, drawing on Westergard, explored the effects that the Devils Hole pupfish was having on the state's administration of water and on stifling the development of agriculture in southern Nevada. He then told readers,

> There is an insecticide on the market called rotenone which has been used successfully to eradicate "problem" fish on many other occasions. This substance holds the key to resolving the "Pupfish Caper" before any more governmental time and money are wasted on this fraudulent attempt to establish federal authority as being greater than Nevada's jurisdiction of its own state water rights. An appropriate quantity of rotenone dumped into that desert sinkhole would effectively and abruptly halt the federal attempt at usurpation.[84]

Rotenone, which in the 1970s was principally used by fishery managers to remove invasive species from waterways, would in Steninger's hands be turned against an endangered one.

The editorial horrified the biologists involved with the pupfish. Joseph Wales, who described *Cyprinodon diabolis* as a unique species in 1930, called the editorial "obviously the product of a mentally defective individual."[85] Acting Region 1 director at the U.S. Fish and Wildlife Service, William H. Meyer, chided Steninger in a letter to the editor, saying that encouraging the use of rotenone was an "irresponsible action which reflects poorly on you and the press in general."[86] In a lengthy and pointed letter to the editor of the *Inyo Register*, which had reprinted portions of the editorial, Phil Pister concluded that "implying the desirability of clandestine violation of State and Federal law" and "attempt[ing] to pervert the minds of his readers represents the term 'insidious' at its very worst."[87]

Steninger did not shrink in front of this and much other criticism. In a follow-up editorial, he doubled down, writing, "It is outrageous, irresponsible, illegal and detrimental to preserve the Devils Hole Pupfish; and it is precisely the opposite to recommend eradication of those 200 little fish." The "usurpation" of state water rights and the supposed theft of "great sums of money and value from Nevada citizens" was by far the greatest evil, pursued only by people in favor of "centralized tyranny."[88] Much as Pister saw the pupfish as an example of the larger issues of environmental protection facing the country, Steninger saw the pupfish as an illustration of governmental overreach.

A few months later, after the Supreme Court's unanimous ruling in support of the principle of a federal reserved water right at Devils Hole, Steninger editorialized that the decision "provided a strong boost for the proposition that Nevada should move to assert state ownership of land being held under federal jurisdiction." Foreshadowing efforts that began a few years later in many western states to reclaim control of federal lands—a movement known as the Sagebrush Rebellion—he concluded, "The only prospect of preserving the kind of freedom and 'blessings of liberty' mentioned in the U.S. Constitution lies in the legislative remedy, with the state legislature taking the initiative to bring the land within Nevada boundaries under Nevada jurisdiction."[89] In 1979, Nevada attempted to do just that when it claimed jurisdiction over 49 million acres of BLM land throughout the state. "Suffice it to say that the Feds do not agree with our position," state attorney general and future governor Richard Bryant told the *New York Times*.[90] Although the Devils Hole pupfish case did not on its own spark the Sagebrush Rebellion, the tensions around the issues of development, employment, and preservation that lurked behind the "kill the pupfish" bumper stickers and Steninger's rotenone argument each found their way into this zeitgeist.[91]

No one actually took up Steninger's call to poison Devils Hole. But if Death Valley National Park records are any indication, breaking into Devils Hole became something of a sport in Nye County. Illegal entry dated back to the moment the Park Service installed a simple gate at the site in 1953, but it seems that destructive entry to Devils Hole increased during the 1970s and 1980s. In 1977, for example, Cappaert's own ranch manager, Barney Barnett, caught three people at Devils Hole who had broken in and destroyed some of James Deacon's monitoring equipment. When Barnett said he was going to call law enforcement, one of the trespassers threw a rock at him. In 1981, James Deacon discovered a spent .30-06 cartridge outside the gate to Devils Hole and the carcasses of two barn owls that had been roosting over Devils Hole.[92]

The motives for these actions, of course, are hard to pin down because none of the perpetrators were caught. In 2016, however, three men drunkenly drove up to Devils Hole in an off-road vehicle, rammed the fence, fired a shotgun at NPS equipment, and one swam in the pool. In contrast to previous incidents, Park Service video cameras captured the incident, and the agency released the footage to the public, asking for help in identifying the culprits.[93] As recounted by the journalist Paige Blankenbuehler in a cover story for *High Country News*, however, the incident seemed to have little or no political content. It was simply the result of a night of drinking that got out of hand, one with lasting consequences for one of the perpetrators who was sentenced to a year in prison for killing an endangered species, destruction of federal property, and for using a firearm after earlier being convicted of a felony.[94]

Even if the record on why particular individuals have broken into Devils Hole is not clear, the broader anger in Nevada toward the pupfish was palpable. Endangered species are often thought of as becoming politically divisive as the result of the power of the 1973 Endangered Species Act. The Devils Hole pupfish, however, did not have the benefits of this statute to protect it through much of the drama with the Cappaert ranch. It nonetheless became one of the first contentious species of the modern environmental era, with similarities to species such as the snail darter and northern spotted owl that each drew protection (and controversy) from the ESA.[95] This may be because, as others have pointed out, fights over species are not really about particular statutes, or even about governmental power, per se.[96] When the federal government used its might to build dams, irrigation projects, and military installations that could fuel growth, Nevadans were more silent about "centralized tyranny." But as the content of federal power changed with respect to land management during the 1970s—to include interpretations that favored other uses of the environment beyond mineral extraction, power production, or farming—the New Deal and Great Society federal government seemed less beneficent.

And while some opposition to the pupfish was certainly opportunistic and cynical—those who had a stake in developing the area further or dreamed of exporting its water to the Las Vegas Valley—it is also true that some people were laid off as a result of the limits placed on pumping around Devils Hole. Neither water law nor endangered species legislation has much of an answer for this. As early as 1968, the head of the endangered species program at BSFW harbored doubts that it would be possible to convince everyone of the importance of endangered species protection, especially in a society with so many other unmet needs. "If you strip away sentiment, morality, conscience, ethics and possibly some religious involvement," he wrote, "I must admit there is little left to convince a big-city baseball fan or a harassed tenement dweller sitting beneath a leaky roof."[97]

But for all these issues that the *Cappaert* litigation could not resolve and the controversy it generated, by bringing the pupfish inside the legal system for apportioning water, it played a crucial role in the pupfish's survival (fig. 5.3). The fact that it was wanted dead explained how it endured. Moreover, *Cappaert* had the effect of creating new partnerships, turning the Nevada Division of Water Resources from an adversary in the pupfish case into a collaborator since, once established by the federal courts, the division had to enforce the Devils Hole water right.

In 1979, William Newman, who had replaced Westergard as state engineer, issued an order "designating" the Amargosa Desert Groundwater Basin, of which Ash Meadows and Devils Hole were a part. The order for the region gave Newman's office additional power to define the preferred uses of water throughout the Amargosa Desert and to limit additional permits for agriculture.[98] A few weeks later, he rejected twenty-eight pending water applications in and around the Amargosa because they "would conflict with and tend to impair the value of existing rights and be detrimental to the public interest and welfare," including the minimum water level set at Devils Hole.[99]

When a farmer north of Devils Hole complained about the designation, Newman replied, "There is no way that all of the patented [privately owned] land in the Amargosa Desert could be irrigated from the ground water reservoir without lowering water tables to the extent that water rights now existing and possibly water levels in Devils Hole would be adversely affected."[100]

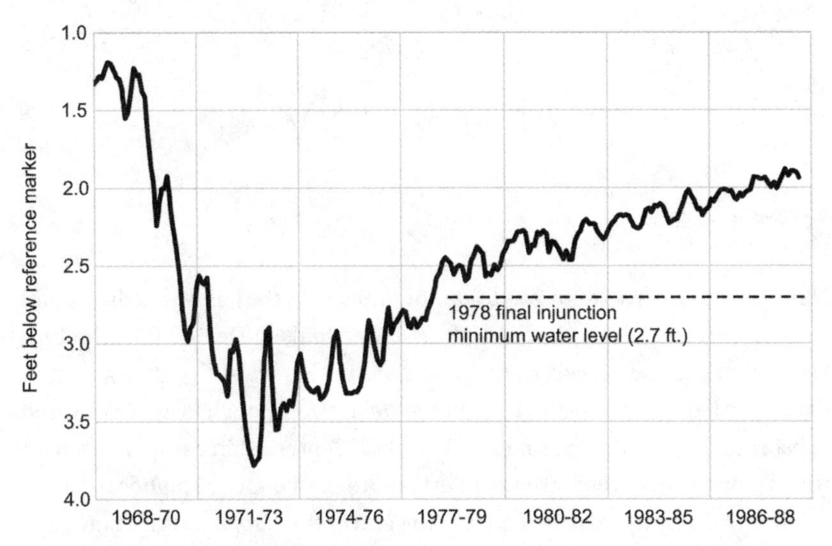

FIGURE 5.3. Lowest daily mean water levels in Devils Hole for each month, January 1968–September 1988. By the 1980s, the water level regularly reached several tenths of a foot above the minimum set by the *Cappaert* litigation. Adapted from *Water-Resources Data for the Devils Hole Area, Nye County, Nevada, July 1978–September 1988*, by Craig L. Westenburg, Open-File Report 90-381 (Carson City, NV: U.S. Geological Survey, 1993), p. 6.

The *Cappaert* case rearranged the relationship between the state and the pupfish and made Newman's own reservations about the effects of limitations protecting the species on other water uses moot. He had to enforce this water right along with all others.

As a result of the pupfish's new relationship to water, one that emerged from the Nevada District Court, the U.S. Supreme Court, and the Nevada Division of Water Resources, the Devils Hole pool rose and stabilized above the level that James Deacon and others thought necessary for its long-term survival. More broadly, these actions over water rights capped a process decades in the making whereby the Devils Hole pupfish endured not only as a result of its evolutionary isolation but also by becoming embedded in several kinds of human institutions. By the end of the 1970s, it seemed that the future was bright for the pupfish. That is, as long as the Park Service and other groups continued to pay attention—and the pupfish behaved.

CHAPTER 6

After Victory

Nineteen ninety-five was a good year for pupfish. In the fall, scuba divers and surface observers for NPS counted 543 individuals in Devils Hole, the fifth year out of the previous six that population surveys registered over 500 pupfish in the habitat. The water level in the Devils Hole pool had also recovered substantially from its 1970s nadir. While still depressed in comparison with the pre-pumping level, it was regularly more than 6 inches higher than the minimum set by the district court at the end of the *Cappaert* litigation. Also in 1995, two backup habitats—concrete tanks in Ash Meadows modeled on the one at Hoover Dam—combined for an estimated population of 215 additional Devils Hole pupfish. In sum, in the mid-1990s, the species seemed to persist, even flourish, as a result of the revised pupfish-human relationship forged in the "save the pupfish" era and embodied in the *Cappaert* decision.[1]

The saga of the 1970s did more than allow the pupfish to survive, it made the species a conservation icon. Press outlets such as the *New York Times* sporadically devoted coverage to the pupfish—including an editorial lauding the Supreme Court for its decision in *Cappaert*.[2] In Nevada, the decision made the pupfish a controversial feature of the state's water and public lands politics. But especially among the biologists and managers who had participated in the drama and their colleagues in university labs and government offices, the pupfish became famous. This remains true in my personal experience talking to people about the species and its story. Biologists and Park Service employees are much more likely to nod with recognition at the words "Devils Hole pupfish" than any other group.

Among the members of the Desert Fishes Council, in particular, the pupfish served as a reminder that it was possible to win sweeping conservation victories for species whose futures once appeared bleak. James Deacon, writing with a coauthor in the early 1990s, remarked that the "legacy of the Devils Hole pupfish" lay in the "symbiotic evolution of species and ethics." The *Cappaert* decision showed that it was possible for society to make choices that would allow a species to continue to thrive, and evolve, in Deacon's words, "undiminished through time." Ultimately, Devils Hole was a success story, one that biologists and conservationists could look to as they faced a host of daunting conservation issues affecting aquatic species across the West.[3]

I wish that I could conclude this book on the same note as Deacon's essay. But the meaning of the stories we tell about conservation (and much else) is heavily dependent on when they end, and the decades after the conservation victories of the 1970s at Devils Hole have been more turbulent and fraught than many would have hoped. Yes, this isolated species persisted through the 1970s as a result of becoming lodged in the ways in which water, land, and power are distributed in Nevada and the West. And the Devils Hole pupfish does remain an inspiring symbol. Perhaps, as Deacon proposed, this survival is even part of a "symbiotic evolution of species and ethics." Events over the subsequent twenty-five years suggest, however, that the actions taken in the 1970s to protect the pupfish were not sufficient to ensure that it could thrive "undiminished through time."[4]

In spring 2006, slightly more than a decade after the Park Service regularly counted more than 400 to 500 pupfish in Devils Hole, divers and surface counters tallied just thirty-eight.[5] While the population is typically lower in the spring than in the fall, the result was astounding: thirty-eight members of the entire species, a new and alarming bottom for a mysterious population decline that began in 1996. The backup refuges, meanwhile, were somehow in worse shape: only one remained operational, and it contained just twenty-six fish that April.[6] Deacon, when asked about the meaning of the pupfish census by the *Las Vegas Review-Journal*, commented, "The best-case scenario is we'll figure out the problem and the population will rebound. Obviously, the worst case is extinction in the next eight months."[7] Instead of a conservation icon—a symbol of persistence—the pupfish was another doomed species.

The reasons for this population crash remain somewhat obscure. Today, scientists have developed two main theories. First, evidence suggests that ecological changes in the composition of algae and invertebrates in Devils Hole have taken place; these shifts may be detrimental to the pupfish's life cycle. Second, research into the pupfish's genome suggests it has very little variability between individuals and that its small population size may be resulting in the buildup of problematic mutations.[8] (A third theory, climate change, is discussed in more detail in the conclusion.) But these explanations were not always self-evident. Thus, uncertainty over the cause of the decline is also a "character" in this chapter. In the 1990s, managers and scientists did not know how to explain the pupfish population decline in part because of decisions that curtailed ecological monitoring of the habitat. Our current ideas about the decline, meanwhile, were generated by studies initiated in response to the problem.

Despite the "worst case" scenarios of 2006 and the difficulty in uncovering the origins of the population crash, the Devils Hole pupfish has survived. The species' population has even recovered to some extent—though it remains well below levels observed in the early 1990s. Even through this episode of near extinction, the persistence of the pupfish depended on a tightening web of connections between the biology, ecology, and natural history of the pupfish—its nature—and the human institutions governing the U.S. West. Tracing these recent events after the victory won in the 1970s, however, does not involve examining titanic changes in science, park management, or water politics but instead reveals the consequences of seemingly small management decisions (or nondecisions). In essence, the nitty-gritty matters when it comes to species survival. Focusing on where research equipment was stored and whether employees made regular monitoring visits to the backup populations, for example, represents the appropriate scale for understanding the species' survival in recent decades.[9]

Following the nuts and bolts of endangered species management in this way reveals one of the most disheartening periods in the modern history of the pupfish—even more than its invisibility within the water rights system that led to groundwater pumping in Ash Meadows in the 1960s. Starting in the 1980s, the Park Service and other agencies charged with protecting the Devils Hole pupfish disinvested in its management, a series of choices that later made it difficult to respond when the pupfish population began its unexplained downward slide in the late 1990s and sometimes put the species at even greater risk. Yet, there is also redemption: NPS and its collaborators were able to change, recovering their own capacity to manage the pupfish for the long run.

Thus, while the experience of the last thirty years makes it hard to conclude the Devils Hole story with the triumph of the early 1990s, the argument stands. The survival of the pupfish was, and is, intimately bound to human history and institutions. To my mind, the post-1995 story illustrates this as well as any other moment in the pupfish's past.

• The Deindustrialization of Devils Hole

Across the United States, but especially in the Northeast, the 1970s and 1980s were characterized by a process called "deindustrialization." Around Pittsburgh, where I attended graduate school and later wrote portions of this manuscript, the region lost 133,000 manufacturing jobs between 1979 and 1987; unemployment peaked above 17 percent.[10] Bruce Springsteen, commenting on this hollowing out in a song about nearby Youngstown, Ohio, noted that steel mill closures turned the land into "scrap and rubble," a feat that Hitler's armies could not accomplish during World War II.[11]

Deindustrialization—a systematic disinvestment in basic productive capacity—can seem far from Devils Hole in both literal and figurative senses. Yet, during that same decade that steel and other heavy industries contracted and made the rust belt, the National Park Service began disinvesting in Devils Hole. This timing was more than coincidence, with Reagan-era economic policy and budgetary pressures on the entire national park system playing a key role in this process. Over the course of the 1980s, the agency reduced its ability to gather ecological knowledge and make reliable management decisions about the pupfish—despite the fact that the outcome of the *Cappaert* case had rested on its investment in hydrological and ecological monitoring. And just as deindustrialization harmed workers and communities built around particular industries, disinvestment by the Park Service eventually hurt the possibilities for effective pupfish management and the species' very survival.

Warning signs for the federal government's lagging commitment to Devils Hole and Ash Meadows emerged in 1977, even before the district court issued its final injunction in *Cappaert*, when the company decided to sell its ranch. The federal government was an obvious potential buyer, since it had just spent the previous seven years defending the Devils Hole ecosystem on the ground and in the courtroom. As mentioned in chapter 5, the Bureau of Sport Fisheries and Wildlife had even studied the possible purchase of Ash Meadows and the creation of a national wildlife refuge there in the early 1970s, but the plan was hamstrung by the Bureau's own hesitancy to make a wildlife refuge for an endangered fish and then by the success of the legal strategy in reducing the threats to the Devils Hole habitat.[12] In 1977, with an even more secure legal settlement, the idea of buying Ash Meadows was no more appealing to the agency. The U.S. Fish and Wildlife Service, the new name for the Bureau after a bureaucratic reorganization in 1974, declined to pursue purchase of the ranch. When James Deacon questioned the wisdom of passing on this opportunity, especially given the other sensitive aquatic habitats within the ranch boundaries, FWS acting regional director, William Meyer, replied, "We would appreciate your rationale for recommending acquisition of the entire property, especially in view of the court's decision on water withdrawals."[13]

Karl Marx once observed that history appears to happen twice, first as tragedy, then as farce.[14] If the Cappaert era was a tragedy in Ash Meadows, then what followed after FWS declined to buy the ranch was certainly the farce. The Preferred Equities Corporation acquired the Cappaert property in 1978. Instead of a base for the production of beef, Preferred Equities envisioned transforming Ash Meadows into "Calvada Lakes," a town with more than 33,000 residential units that would result in a city of some 75,000 people.[15] (At the time, Nevada's state capital, Carson City, had just 32,000 residents.[16]) Such a plan for Ash Meadows represented a serious threat not only to the Devils Hole pupfish but the three other unique fishes in Ash Meadows, as well as numerous endemic plants and invertebrates. To return to a metaphor used at the start of this book, Ash Meadows can be understood as an island of biodiversity, the Galápagos of the Intermountain West. Under Preferred Equities' ownership, the Galápagos would also become a sprawling desert suburbia in the wide orbit of Las Vegas, complete with cul-de-sacs, parking lots, and shopping centers.[17]

Part of what made this plan for Ash Meadows farcical was that the water source for this development remained unclear. While the purchase of the ranch came with the surface and groundwater water rights held by Cappaert, these were limited by the court decision. A report to Nevada's attorney general in 1982 suggested that even "by the most liberal estimates" the company would only be able to meet the demand for water from 8.5 percent of the "projected number of people this project is supposed to handle."[18] Nor did Preferred Equities have a particularly scrupulous reputation as a developer by the time it acquired the Ash Meadows land. In the early 1970s, the *New York Times* published an exposé of the company's high-pressure sales pitch for another development in nearby Pahrump, Nevada, where the company pushed tourists into purchasing building lots as hot investments.[19] The company's owner had also been in trouble with authorities in Florida (for selling land that turned out to be underwater) and the federal government (for tax fraud and swindling a group of West German investors).[20]

The Fish and Wildlife Service was able to recover from its serious misstep in declining to purchase Ash Meadows from Cappaert—a feat that separates this episode from others taken later in the 1980s at Devils Hole. In the face of the Calvada Lakes threat to the Ash Meadows ecosystem, an unusual group assembled and managed to scuttle the effort to build the subdivision. In the process, it redeemed a wish of Carl Hubbs from the 1940s: that the springs across Ash Meadows should be permanently protected to preserve their unique aquatic life. (Hubbs did not live to see this; he passed away in 1979 at the age of 84.) Biologists at FWS came to see the decision not to purchase Cappaert's holding in 1977 as a serious blunder—a "flagrant error in judgment by the Service," one supervisor wrote. They closely monitored the effects of Preferred Equities' development, eventually listing on an emergency basis two Ash Meadows fishes (the Ash Meadows Amargosa pupfish and the Ash Meadows speckled dace) through the Endangered Species Act.[21] Once listed, any development that might result in the "take" of these fishes in Ash Meadows springs could also result in legal action by FWS.

While these actions could slow or halt the construction of Calvada Lakes, the Endangered Species Act does not require the purchase of lands for conservation. That is, the new listings under the Endangered Species Act would not automatically result in a wildlife refuge. Resolving the situation favorably for Ash Meadows species required the diligent work of a then-growing force in the conservation: private land trusts. Field staff for one such group, the Nature Conservancy, helped push Preferred Equities to see the wisdom of willingly selling its lands.[22] At the same time, they worked with FWS and Nevada Senator Paul Laxalt—a "sagebrush rebel"—to secure funding to pay for a refuge.

By 1983, the Nature Conservancy arranged for Preferred Equities to sell the organization its land for $5.5 million, with the federal government reimbursing the Conservancy using funds from the Land and Water Conservation Fund, established in the 1960s to provide for conservation projects through the use of fees from off-shore oil drilling projects in federal waters.[23] And in the summer of 1984, Ash Meadows became the nation's 422nd national wildlife refuge, one of the first established specifically for the protection of endangered species.[24] Signaling the broadening interests of FWS, Ash Meadows' lack of migratory waterfowl, which had deterred the agency in the 1940s from acquisition, did not play a role in the story.

Acquiring Ash Meadows, however, was an exception to the rule in federal land management in the early 1980s. The attitude of Ronald Reagan's first Secretary of the Interior, James Watt, was decidedly against federal land acquisition. Watt arrived in Washington in 1981 telling the press, "America's resources were put here for the enjoyment of the people, now and in the future, and should not be denied to the people by elitist groups," a framing that could be pressed into service against the scientists, managers, and conservationists looking to protect Ash Meadows from the ill effects of development.[25] The Reagan administration also called for ending land acquisitions with the Land and Water Conservation Fund.[26] Although Watt was unable to entirely halt appropriations from the fund, during his tenure its use dropped precipitously from Carter-era levels.[27] Ash Meadows bucked this trend as a result of the dire threat that the Calvada Lakes development posed and because Preferred Equities' willing sale likely avoided litigation that would have held up the development in court. Paul Laxalt, the Nevada senator who helped make the acquisition a reality, was once quoted in the *New York Times* saying that Nevadans were tired of "being ruled like some faraway colony by uncaring and unknowledgeable bureaucrats" in Washington.[28] Yet he defended his support for the wildlife refuge to a constituent, writing that while the purchase was "contrary to my long-held position that the Federal Government should not acquire additional lands[,]...the situation at Ash Meadows is unique and critical."[29]

The establishment of Ash Meadows National Wildlife Refuge might have kickstarted a renaissance in monitoring and research at Devils Hole. With its creation, three agencies had overlapping responsibilities to the Devils Hole pupfish—a situation that continues to the present day. The U.S. Fish and Wildlife Service had obligations stemming both from its role as the administrator of the Endangered Species Act and because the Ash Meadows refuge that it managed played home to two concrete backup tanks for Devils Hole pupfish. The Nevada Department of Fish and Game—renamed the Nevada Department of Wildlife (NDOW) in the late 1970s—was charged with conserving all wildlife in the state and oversaw the backup refuge tank at Hoover Dam. The National Park Service functioned as the de facto leader of this triumvirate during the 1980s and 1990s, since it was responsible for managing Devils Hole as part of the national park system.

Instead of investigating new ways to cooperate at Devils Hole, however, the Park Service spent the years after the establishment of Ash Meadows National Wildlife Refuge stepping away from the threads of history that bound it to the pupfish. Before the ink was even dry on Preferred Equities' sale of its Ash Meadows property, Park Service employees began discussing how to get out of the pupfish business by turning management over to FWS. One NPS planner wrote that Devils Hole would be better in the hands of FWS since they were dedicated to "management," while the Park Service to "natural processes."[30] The Death Valley draft general management plan, published in 1988, concurred.[31]

For the preceding fifteen years, the Park Service clearly demonstrated that it *was* capable of management at Devils Hole. Yet, the lack of enthusiasm for the Park Service's role at Devils Hole reflected the long-lasting power of the vision offered by agency leaders in 1950 who agreed on the importance of Devils Hole as an object of scientific inquiry but believed protecting such a site fell outside of NPS responsibility.[32] Ultimately, Devils Hole remained part of Death Valley National Monument, perhaps because FWS was not eager to take sole charge. "Quite frankly," one Death Valley employee commented, "the National Park Service is capable of committing more funding, personnel, supplies and equipment, to Devils Hole than can USFWS."[33] Death Valley management might not have agreed. Indeed, one of the reasons that transferring Devils Hole to FWS was appealing to NPS managers in Death Valley had little to do with the differing missions of the two agencies and everything to do with funding.

The Park Service's relationship with the pupfish was—in legal terms—governed both by Devils Hole's addition to Death Valley under the purview of the Antiquities Act and by the agency's interpretation of its mission, as spelled out in its own Organic Act. Yet neither of these authorities—nor later legislation such as the Endangered Species Act—required the Park Service to dedicate specific funding to the pupfish. As it faced financial pressure during the 1980s, Death Valley management used its administrative discretion to simply defund the monitoring that had formed the basis of the government's case in *Cappaert v. United States*. The defunding suggested that some of the ties between people and the pupfish, which had enabled the species' survival, were not inevitable or irreversible.

Beginning in 1972, as the habitat in Devils Hole shrank rapidly, the cornerstone of pupfish monitoring was population censusing conducted by scuba dives. Since 1974, James Deacon from UNLV had been the Park Service contractor in charge of these population counts. But as we have seen, counting pupfish was only part of his work at Devils Hole in the 1970s; he also collected data on sunlight, dissolved oxygen, phosphorus, and nitrogen, as well as making observations on algal growth, the role of windblown material in the habitat, and the effect of earthquakes.[34] This research undergirded the federal government's legal argument for a minimum water level of 2.7 feet below the copper reference marker in Devils Hole.

After the court case concluded, Deacon continued his work for NPS, monitoring the population through monthly scuba population censusing. By 1984, he proposed undertaking more intensive analyses of the ecological interactions in Devils Hole. Reducing the number of scheduled dives, Deacon told the monument, would provide time to "examine the relationship between nutrients, algal growth and population size," as well as to explore the size and sex ratio of pupfish in Devils Hole in order to better compare them with populations in the artificial refuges.[35] Death Valley resource staff supported shifting the research plan during 1984, with one manager (a former Desert Fishes Council president) summarizing Deacon's idea by telling superintendent Edward Rothfuss that "pupfish population censuses through the years have quantified cyclic and non-cyclic fluctuations. What is now needed is to learn what causes fluctuations—especially noncyclic ones."[36]

Carrying out such monitoring and research was also well-supported by the community of managers and researchers concerned with the pupfish. The 1980 "Recovery Plan" for the Devils Hole pupfish, produced by an FWS-organized team that Deacon headed, had recommended much the same thing: continual ecological monitoring "whereby changes in biological, physical, and chemical parameters can be detected and rectified if deemed necessary." The plan specifically mentioned the monitoring of "food items, algae and invertebrate fauna."[37]

Ideally, recovery plans are created for every species listed under the Endangered Species Act and are meant to compile the best information on actions that will improve the condition of the species. They are not, however, binding documents. Despite the importance of this research—especially for understanding the causes of "noncyclic" changes in the pupfish population—the work had barely begun when Deacon's contract was unceremoniously terminated at the end of 1985, saving Death Valley $4,000 per year (about $10,000 in 2020 dollars).[38]

Budgetary pressures at Death Valley and a lack of institutional commitment to ecological monitoring and management made the money destined for Devils Hole an easy target. During the 1980s, Death Valley's base budget did not grow much, and additional funding for special projects fluctuated wildly, meaning that, as historians Hal Rothman and Char Miller observed, "it became difficult for the park to determine how to deploy its inconsistent resources."[39] At the same time, across the park system, NPS did little during the 1980s to build its capacity to inventory and monitor park resources.[40] A 1987 Death Valley superintendent's report tried to frame the Devils Hole budget cut as planned, claiming that Deacon had "completed his five-year NPS-funded study." The work was, however, decidedly incomplete and was meant to continue according to the Devils Hole pupfish recovery plan and earlier monument budget planning documents.[41] The Devils Hole pupfish survived Cappaert Enterprises, in part, because of ecological monitoring. The Park Service's decision implied it could survive without it.

There is a caveat to Death Valley National Monument's actions in the 1980s. NPS did not entirely turn its back on Devils Hole. After the end of monument-funded monitoring, semiannual pupfish population counting did continue through the development of a volunteer dive team that worked with NPS staff to ensure that some data on the pupfish's population would be gathered.[42] And the Park Service continued to pay attention to the regional groundwater situation, especially after the Las Vegas Valley Water District applied to the state in 1989 for groundwater permits from basins around southern Nevada, some of which were hundreds of miles away. (Thirty years after the initial applications were filed, this situation remains contentious and unresolved.[43]) Yet pupfish counting and water monitoring did little to develop additional data on the Devils Hole ecosystem or help answer questions about the relationship between pupfish populations, algae, and nutrients there, especially the noncyclic ones Deacon discussed in 1984. The reduction in monitoring also put a lot of pressure on the data that the agencies did accumulate: if the population began to decline, would its cause be easily identifiable?

As early as 1976, James Deacon wrote that the Park Service should "establish administrative procedures that recognize the value of the Devils

Hole pupfish which can only be preserved if we learn enough about it to effect preservation—or failing that, only if we are lucky."[44] The consequences of cutting the ecological monitoring at Devils Hole meant that the Park Service essentially counted on being lucky. As the Park Service disinvested in Devils Hole, the agency walked back from some of the enmeshment with human institutions and practices that had allowed the pupfish to survive the first eight decades of the twentieth century, especially the years after 1968. And while its decision did not *cause* the population decline of Devils Hole pupfish that began in the mid-1990s, once underway, the trio of agencies were left flat-footed as they sought to understand its causes and formulate responses.

• Luck Runs Out

Death Valley National Park (redesignated to "Park" status after the passage of the California Desert Protection Act in 1994) organized a semiannual pupfish population survey at Devils Hole on October 26, 1996, conducted by volunteers and agency staff. The group counted 433 fish in the habitat. In distributing the information, Death Valley National Park wildlife biologist Doug Threloff noted this result "appeared to be significantly lower than counts which were conducted in past years." The pupfish seemed to be behaving normally, however, and Threloff added that future counts "should provide a more definite answer as to whether anything significant is happening to affect the number of pupfish present."[45] Waiting on the next population survey was about the only choice that the Park Service had, since, aside from the water level data, no other aspects of the habitat were being regularly monitored.

Over the next several years, Threloff's concern proved correct: the pupfish population in Devils Hole all but collapsed (fig. 6.1). Yet in contrast to biologists' rapid response in 1969 and 1970 to the declining water level in Devils Hole, wherein they organized a symposium, advocated successfully for the establishment of an Interior Department interagency task force, and then pushed for a large, cooperatively funded groundwater study—all in less than twelve months—it took years for the Park Service, the U.S. Fish and Wildlife Service, and the Nevada Department of Wildlife to recognize and mobilize a coordinated response to the problem of declining populations after 1996.

Part of the reason for a muted response lay in the obscure and slow-motion nature of the crisis, in comparison to rapidly declining water levels

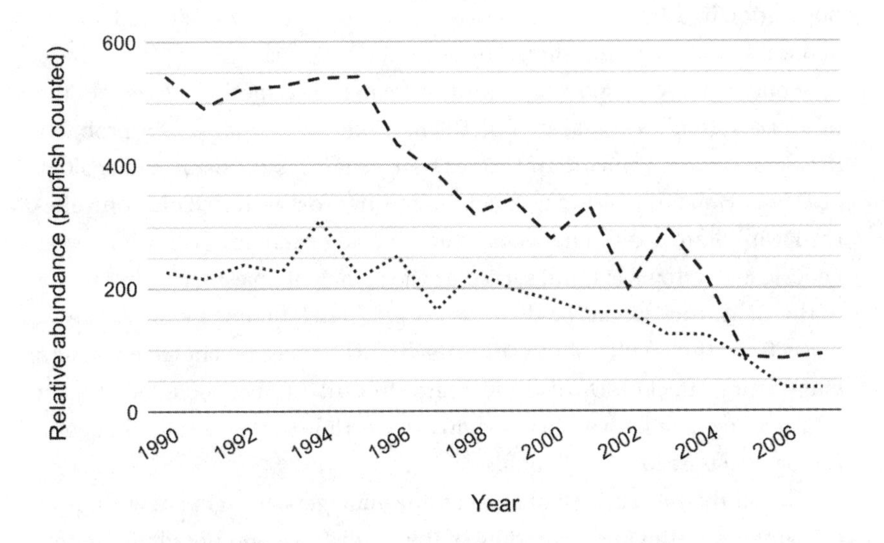

FIGURE 6.1. Surveys revealed that the Devils Hole pupfish population collapsed between 1996 and 2006. The top line in this chart shows the maximum number of pupfish counted in Devils Hole each year (usually in the fall), and the bottom line shows the minimum counted (usually in the spring). Biologists refer to these population estimates as describing the "relative abundance" of pupfish in Devils Hole, since the numbers include only those fish observed by the counters, not every individual pupfish in the habitat. Chart by author with data from National Park Service.

in Devils Hole set off by groundwater pumping in 1969. At first, even James Deacon doubted that the declining population data meant much; the 1996 data could easily have resulted from fewer pupfish being observed, rather than from fewer pupfish in the habitat.[46] But beyond the different particulars, the lethargic response of the agencies in the late 1990s was also an outgrowth of the disinvestment in Devils Hole during the previous decade.

In April 1998, eighteen months after the troubling census results, Doug Threloff and Death Valley organized and hosted a Devils Hole Biological Workshop.[47] The meeting took place west of Ash Meadows at Longstreet Casino on Nevada Highway 373—the closest place to Devils Hole for placing a bet or ordering a cheeseburger. Over two days, the group formulated a plan to correct "current data deficiencies" at Devils Hole, notably by commissioning a study of "bioenergetics"—basically, mapping the Devils Hole food web. The plan also raised the possibility of a long-term ecosystem monitoring effort.[48] In essence, this meeting responded to the declining population by admitting that there was not enough data to make judgments about its causes or what should be done about them.

Funded by NPS and USGS, a three-year bioenergetics study was the first ecosystem-level research at Devils Hole of the 1990s, and probably the most detailed since Carol James's 1969 master's thesis on the ecology of Devils Hole. The study was conducted by Northern Arizona University's Dean Blinn, a veteran investigator of ecological relationships in desert aquatic ecosystems.[49] Blinn served as the principal investigator, but much of the field work would be done by his graduate student Kevin Wilson.[50] Aside from the results of the study itself, this personnel choice has had a long-lasting effect on Devils Hole. Years after this study concluded, Wilson returned to play a key role in managing Devils Hole as Death Valley's aquatic ecologist, a position he still holds.

The study was a breath of fresh air for managers, bringing new kinds of information to the understanding of the pupfish beyond population numbers.[51] The study even found some things that might be troubling for the pupfish. At a 2000 meeting with the Park Service where they shared their initial findings, Blinn and Wilson showed that the water temperature on the shallow shelf spiked periodically, pushing it to the point where it might interfere with egg development. They also found that a planaria (flatworm) in the pool might be predating pupfish eggs. Finally, they noted that gravel deposition on the southernmost part of the shallow shelf caused this portion to routinely go dry at low tides. (Devils Hole is so deep it experiences tides.) This reduction in space for spawning, egg development, and foraging might simply be making Devils Hole a smaller habitat as far as the pupfish were concerned, resulting in a lower population.[52]

But this information raised nearly as many questions as it answered. Without historical data for comparison, it remained difficult for the researchers or managers to understand whether these findings represented the norm for Devils Hole or a departure that explained the declining pupfish population. After Blinn and Wilson shared their first findings, in December 2000, Doug Threloff penned a long memorandum to the superintendent at Death Valley that addressed precisely this uncertainty.

> One of the most vexing dilemmas related to the decline of the pupfish is that there is a limited amount of baseline ecological data that was collected prior to the fish decline. Park staff do not therefore

> have quantitative data that could be used to determine how the ecosystem has changed over time. If "vital signs" datasets for critical aspects of the Devils Hole ecosystem had been collected when pupfish numbers were at optimum levels in the early 1990s, it would be easier to compare the current environment with low fish numbers to a "healthy" system with high fish numbers.[53]

Years later, when I interviewed Threloff, he expressed regret at the "big mistake" of not fighting for a monitoring program before the population began to tumble, saying he "really should have... had enough forethought to launch a good, intensive effort to understand what really drove Devils Hole when the fish numbers were high."[54] At the time of our conversation, I had not pieced together the story of Death Valley's disinvestment in the 1980s and never asked him whether he knew that Deacon had planned just such a monitoring strategy until his contract was eliminated, years before Threloff arrived at the park.

Threloff's letter represented a pivot point in management of the pupfish, and not only because he used it to resign from his post in Death Valley (he had accepted a position in Ventura, California, with the Fish and Wildlife Service). Between 1996 and 2000, the Park Service had focused on improving the knowledge about the Devils Hole pupfish, principally through Blinn and Wilson's ecological study. But the continuing population decline began to put pressure on the agencies to take actions to stem or reverse the emerging trend. In other words, they increasingly faced the question of what they could do to "save the pupfish." Over the next several years, the continued information deficit, overstretched staff, and lack of clear management authority produced confusion and several self-inflicted wounds that made this question difficult to answer and the pupfish's survival more precarious.

In late 2002, to improve its prospects, FWS created a new "recovery team" for the Devils Hole pupfish. The idea for such teams stretched back to a 1978 amendment to the Endangered Species Act that called for developing recovery plans for each listed species.[55] From the outset, it was unclear what exactly the new Devils Hole pupfish team would accomplish and what power it had, especially since the pupfish's own recovery plan had been published more than two decades earlier (and the agencies had spent about the same number of years ignoring its recommendations). One FWS memo described the new team as a vehicle to "develop and implement actions necessary for recovery and management" of the pupfish.[56] In contrast, meeting notes produced by FWS later called the team a forum to "discuss and provide recommendations to the Fish and Wildlife Service regarding activities that may impact Devils Hole pupfish."[57] One characterization made the recovery team a collaborative decision-making body, while the other vision placed the recovery team's role squarely in an advisory capacity to FWS. Neither formulation, meanwhile, clarified the terms under which cooperation between the Fish and Wildlife Service, the Park Service, and the Nevada Department of Wildlife would occur.

Beyond mixed signals about the group's purpose and authority, the recovery team faced a truly overwhelming list of tasks—essentially a backlog of projects and needs that had been continually deferred or left undone since the 1980s. At the first recovery team meetings in 2002 and 2003, attendees from agencies and universities discussed a dizzying range of issues. Just a cross section from the agendas and notes from these gatherings includes such topics as (1) exploring the possibilities for captive propagation (aquarium rearing) of pupfish; (2) a proposal to write a new pupfish census scuba dive protocol; (3) whether the current absence of an ostracod (a small crustacean) in Devils Hole that was present in the 1960s might explain the decline in pupfish populations; (4) the need to compile all existing data on Devils Hole in one place; (5) how to respond to a water-intensive solar-power development then planned for the Amargosa Valley; (6) evaluating sediment composition on the shallow shelf; (7) creating short- and long-term monitoring protocols; and (8) writing a new management plan for the captive, backup populations.[58]

These issues, and many others, existed at different spatial scales, required different kinds of expertise, and had distinct time horizons for their respective resolution or implementation. For a group of agencies with no employees dedicated to Devils Hole on a full-time basis, moving forward on these tasks must have seemed daunting, especially as pupfish population counts continued to signal bad news. (A January 2003 survey observed just 124 pupfish, the lowest number since the counting began in 1972.)

The issues raised at the first recovery team meetings were so challenging, it seems, that parts of the "recovery process" simply stalled. At one point in spring 2004, an NPS Water Resources Division employee sent an email to the recovery team coordinator at FWS asking, "When is the next meeting of the Devils Hole Recovery Team? It has been a long time since the group has met."[59]

Beyond struggling with new tasks that would enable pupfish persistence, the agencies faced problems just maintaining existing projects. The fate of the artificial backup refuges built to prevent the extinction of the pupfish demonstrated this problem. In addition to the Hoover Dam Refuge (first stocked with pupfish in 1972), two other, nearly identical, concrete habitats for Devils Hole pupfish were later built in Ash Meadows and utilized for backup pupfish populations: at School Springs, stocked in 1980, and at Point of Rocks, stocked in 1991. All three refuges faced perpetual neglect. The 1980 Devils Hole pupfish recovery plan, for example, stipulated that a small number of fish should be transferred annually from Devils Hole to each operating refuge because of concerns that the genomes of these refuge populations would begin to stray away from the main pupfish population.[60] Despite this goal, managers only moved fish to the refuges from Devils Hole four times between 1980 and 1995.[61]

But genetic problems did not cause the population's decline. Between 2003 and 2006, all three of these refuges failed in their role as backup habitats for prosaic, preventable reasons. Such failures made the pupfish more vulnerable to extinction. At the School Springs refuge, what ultimately killed the population was a faulty water delivery system and irregular checks of the population by Ash Meadows National Wildlife Refuge staff. In 1999, an improperly adjusted valve prevented warm water from reaching the tank, dropping the habitat's temperature and killing an estimated seventy pupfish before the situation was corrected.[62] After the accident, Ash Meadows staff produced a new management plan stipulating that it should be visited daily in order to assure its proper operation.[63] Daily inspections never became regular, however, and in 2002 a pump failure went unobserved for at least eight days and led to the temperature in the tank plummeting from 91°F to 62°F, reducing the population. After operating continuously since 1980, the last fish in the refuge tank was sighted in January 2003.[64]

The Hoover Dam and Point of Rocks refuges suffered different fates, though they were equally tied to management problems. At the Hoover Dam refuge, employees noticed in 2006 that the tank had been colonized by an invasive snail that is often accidentally transported to new habitats by fishing gear. In this case, however, no anglers visit the habitat; the likely method for introduction was contaminated research equipment. An explosion in the snail population—it coated all surfaces of the tank—made the habitat unsuitable and all remaining pupfish were eventually moved to aquaria where they died without producing a viable next generation.[65] At Point of Rocks, meanwhile, in 2005 biologists realized that the Devils Hole pupfish population in the concrete tank had become hybridized with its close genetic relative, the Ash Meadows Amargosa pupfish.[66] (Hybridization is generally proscribed for endangered species. Since both the Devils Hole and Ash Meadows Amargosa pupfishes are listed species, the policy is to keep them isolated from one another, lest their unique characters be compromised.[67]) Much speculation has focused on exactly how the Ash Meadows Amargosa pupfish made it into the concrete tank. Leafing through the logbooks of Ash Meadows staff visits to the tank, I found that in 1996 two employees discovered three fish swimming just outside the tank (in the "outflow box") and deposited them into the tank, apparently believing them to be escaped Devils Hole pupfish. Instead, however, they may have been Ash Meadows Amargosa pupfish that swam upstream.[68] Collectively, in a period of a few short years, a longstanding lack of adequate oversight led to the failure of the backup refuges, just when they might become necessary for preserving the species.

The refuges were not the only place where management failures occurred. Even when the Recovery Team did take proactive steps to learn about the pupfish—such as authorizing Southern Oregon University researchers to study larval fish in Devils Hole—it could end in disaster. On September 11, 2004, an intense storm caused a flash flood at Devils Hole, the second such event in less than a month.[69] Along with the rainwater, sediment, and rocks that rushed toward the pool, a large plastic crate containing thirty-six glass jars fitted with plastic funnels became dislodged from its storage place behind a rock and tumbled toward the water. The traps had been used to sample the abundance of larval pupfish on the shallow shelf by allowing for their temporary capture.[70] Swept into the habitat, however, these jars and funnels began trapping adult pupfish, and over twelve days killed approximately eighty, probably more than a third of the entire adult population in Devils Hole (fig. 6.2).

The tragedy stemmed not only from bad luck, but as with the artificial refuges, from lax oversight. First, NPS had approved the storage of the equipment in Devils Hole, despite a well-known history of flash floods. Second,

FIGURE 6.2. A National Park Service employee holds a jar with decaying Devils Hole pupfish after a flash flood washed improperly stored research equipment into Devils Hole, killing more than seventy fish, September 2004. Photograph courtesy National Park Service.

more than a week passed before any Death Valley staff visited Devils Hole, and then the unfolding tragedy did not become apparent for another two days.[71] FWS staff from Ash Meadows were, of course, in the area during this time, but no arrangement with NPS existed for them to check on Devils Hole.[72] Afterward, the agencies never even produced a press release about the event, and the public did not learn about it until nine months later, in June 2005.[73] The flood may have been an act of God, but the pupfish deaths resulted from human actions.

The 2004 fish trap incident represented a new low point for pupfish management in the post-Cappaert era. Alongside the collapse of all three concrete refuge tanks between 2003 and 2006, the tragedy showed how the unraveling of the monitoring practices and the decline in administrative capacity at Devils Hole could lead the species to the brink of extinction. The pupfish had survived the budget cuts of the 1980s and the tepid agency response to declining population numbers in the 1990s, but it was unlikely to survive indefinitely without reforging the relationships that had delivered the species through the twentieth century.

The Fire Line

Starting in 2005, and especially after the spring 2006 pupfish count revealed fewer than forty fish in Devils Hole, the Park Service, Fish and Wildlife Service, and NDOW made halting and frantic efforts to rescue the pupfish from extinction. The agencies removed sediment from the shallow shelf that appeared to be interfering with successful reproduction and engineered a complex series of transplants to aquaria from both Devils Hole and the failing Hoover Dam Refuge in last-ditch efforts to try raising pupfish in captivity—something never successfully done, despite intermittent efforts stretching back to the 1930s. The aquaria attempts again failed.[74]

The managing agencies were hamstrung by more than the real difficulties of reversing the pupfish population decline or successful captive propagation. Since the creation of the pupfish recovery team in 2002, if not before, the lack of a clear chain of command and accountability had haunted and frustrated the trio of agencies, resulting in petty defensiveness as well as incomplete projects. Any effort to "save the pupfish" would have to address both. At one point, the Park Service and the Fish and Wildlife Service could not even agree on how and when to access water monitoring equipment on the shallow shelf of Devils Hole in a way that would be safe for humans and pupfish alike.[75] More broadly, Death Valley staff had the feeling that while the Fish and Wildlife Service had been negligent on moving forward with recovery team proposals, it still wanted to be in the driver's seat of pupfish management because of its role in enforcing the Endangered Species Act.[76]

Assigning blame, however, would not change a climate that was clearly unconducive for making successful short-term and long-term management decisions. At Devils Hole, Mike Bower, a recent addition to the Death Valley staff dedicated to Devils Hole, was a perceptive analyst of the situation he had walked into. Writing to his supervisor and Death Valley superintendent J. T. Reynolds, Bower reported that a "lack of clear roles, responsibilities, and authorities has cast a cloud over much of our effort.... I am not sure if we [NPS] are a threat, if there are fragile egos, if it is agency culture, or if it is simply a lack of respect (on our part or theirs [FWS]). Regardless, I hope we can find a productive way to assign our human resources to optimize our chances of recovering this fish."[77] The agencies needed a way to manage themselves as much as they needed a way to save the pupfish.

There is nothing new under the sun. And the Park Service, Fish and

Wildlife Service, and Nevada Department of Wildlife staff responsible for the pupfish were not the first group of public employees to have difficulty working together. In fact, the problem of how to facilitate cooperation across organizations is a widespread concern in land management and endangered species preservation. (The first chapter of an academic book on this issue begins with the observation that the federal system of government was "designed to prevent tyranny, not to achieve policy coherence."[78]) Yet because the problem pupfish managers faced was not new, they were able to find guidance from elsewhere in government: wildland fire management.

Over a brief two-week period in 1970—they same year that the Interior Department banded together in the interagency pupfish task force—seventeen major fires burned 500,000 acres, destroyed 700 structures, and resulted in sixteen fatalities in southern California. As the fires crossed jurisdictional boundaries—from a national forest to private land—it required cooperative responses from multiple agencies: the U.S. Forest Service, the California Division of Forestry (now known as CALFIRE), and local fire departments. These agencies discovered that there was no centralized source for accurate and current information on fire location or behavior and that units from different agencies even had difficulty communicating directly since they used different radio frequencies. Once they could talk to each other, meanwhile, firefighters found they had different terminology, procedures, and organizational structures.[79] Cooperating under these conditions was challenging and dangerous.

In the aftermath of this fire season, the U.S. Forest Service undertook a multiyear study it titled FIRESCOPE to improve how firefighting agencies cooperated. One of the central outcomes was something called the "Incident Command System," or ICS. Especially important in the new system was the idea of a clear unity of command that crossed typical agency boundaries, allowing for the coordinated deployment of fire crews, engines, and air resources, all using compatible terminology and equipment. The Incident Command System remains a cornerstone of wildland firefighting and is now a key component of the Federal Emergency Management Agency's National Incident Management System.

In 2006—after divers counted just thirty-eight pupfish in the habitat—this system for fighting wildfires and responding to emergencies was applied to the pupfish, the first and only time that I have found ICS being used in endangered species management. "The Devils Hole pupfish recovery effort is now time-critical, and would benefit from a more structured organization such as the Incident Command System," Death Valley informed the collaborating agencies in July.[80] (It was no accident that Death Valley Superintendent Reynolds played a critical role in bringing ICS to pupfish management. Reynolds was a veteran NPS employee and had served in a variety of law enforcement, search-and-rescue, and firefighting roles in national parks since the early 1970s.[81]) At the heart of this change was a system that would allow "a better clarification of roles and responsibilities, which in turn would promote positive forward action," and provide mechanisms for resolving impasses—exactly what the recovery process had lacked since its organization in 2002.[82] Although agencies could—and did—still disagree about a range of issues, this structure ensured that such disagreement did not result in inertia or confusion over authority, as it had earlier in the decade.

Still, the Incident Command System was not a silver bullet for pupfish preservation. The first major management action after the adoption of this structure failed: an attempt to restart a backup refuge. After the fall 2006 pupfish count revealed eighty-five individuals, leaders from all three agencies agreed to an NPS proposal to transplant pupfish from Devils Hole to Point of Rocks Refuge, where all the hybridized pupfish had been already removed. The action, they hoped, would result in a new, self-sustaining backup population that could also serve as a basis for additional aquaria propagation efforts. In December, twelve fish were moved to the Point of Rocks Refuge. By the following month, however, a snorkel survey revealed just one female and one male Devils Hole pupfish—ten had apparently died within weeks of the transfer; ultimately the plan failed.[83]

But if the Incident Command System could not guarantee success, it did facilitate consensus on proposals and risks worth taking. During the fish collection for the Point of Rocks refuge transplant in December 2006, biologists noticed that "the fish [in Devils Hole] appeared to be thin, coloration was pale and mottled and there was erosion of the margins of the dorsal and caudal fins."[84] Although there was not any historical data with which to compare their condition, managers believed the fish could be malnourished, a belief based not only on the physical condition of the fish but also a sense that the Devils Hole ecosystem had become a less hospitable place for pupfish. In response, on January 4, 2007, the agencies began a supplemental feeding program by regularly adding commercial fish food to the habitat.[85]

Over the next months, the feeding continued; with modifications, it remains in place to the present day. Such an action would not have been possible just a year earlier, since the agencies would have been unable to reach consensus without the Incident Command System. And by taking the step, the managers had also formed a new, intimate, relationship with the species to enable its survival.

The agencies could move forward with such projects not only because of a new organizational structure but also because of increased funding and staffing. By 2008, three full-time, permanent personnel had been hired at Death Valley National Park to manage aquatic ecosystems, principally Devils Hole.[86] With the support of the Fish and Wildlife Service and the Nevada Department of Wildlife, the park finally wrote and implemented a long-term ecological monitoring plan, the idea for which stretched back to James Deacon in the 1980s and had been recommended periodically by managers and scientists. Kevin Wilson, the graduate student who had worked on the bioenergetics study starting in 1999 and was hired by the Park Service after he completed his doctorate, wrote the plan.[87] The park also used its staff to upgrade the infrastructure at Devils Hole, including improved fencing, research and monitoring access to the shallow shelf, and site security.[88] And, along with the Fish and Wildlife Service and the Nevada Department of Wildlife, the Park Service encouraged, facilitated, and funded new university-based research on the pupfish that increased knowledge on a wide range of issues connected to Devils Hole.[89] Each of these ideas had been discussed in some form for years but only bloomed after the implementation of the Incident Command System and a significant investment in personnel.

Not to be outdone, the Fish and Wildlife Service made its own substantial and enduring investment in the pupfish's survival, constructing and staffing a new artificial refuge in Ash Meadows. The Ash Meadows Fish Conservation Facility (FCF) reimagined the idea of an artificial refuge in much more ambitious and complex ways than the variations on the concrete 1970s-era Hoover Dam design previously replicated in Ash Meadows at Point of Rocks and School Springs. The idea for a new refuge stretched back to the recovery team discussions in 2003, but it was not completed until 2012.[90]

The finished FCF is an impressive facility located on the site of the old School Springs Refuge. The building—it is the closest structure to Devils

Hole, less than a mile away—looks like a futuristic barn compound from a *Mad Max*-type desert dystopia and features a 100,000-gallon pool (approximately ten times larger than the earlier concrete refuges) that mimics the shape, depth and lighting of Devils Hole, including a replica of the shallow shelf area. (Journalist Elizabeth Kolbert has dubbed the facility "Devils Hole Jr."[91]) One room adjacent to the tank is an elaborate computer-controlled pumphouse that allows the staff to fine tune temperature and water chemistry. Also housed at the FCF is a wet lab containing shelves of aquarium tanks for raising pupfish eggs removed from Devils Hole. The fish facility is in some ways the photographic negative of the supplemental feeding program in Devils Hole. In the fish facility, the Fish and Wildlife Service has gone to incredible lengths to build a refuge that is as Devils Hole–like as possible, while in Devils Hole itself, the Park Service has worked to change Devils Hole so that it is more hospitable to the pupfish.

The source of funding for the refuge—it cost $4.5 million to build—is itself a reminder of the ways in which the political and economic geography of Nevada has become intertwined with the pupfish.[92] In 1998, Congress passed the Southern Nevada Public Land Management Act, known by the not-so-catchy acronym SNPLMA (*snip-la-ma*). The act provided for the sale to developers of federal land in the Las Vegas Valley, helping fuel the 2000s boom in construction—and water usage—there. Over decades, the demand for water that this legislation helped create may have long-term effects on the groundwater that supports springs around southern Nevada, but it also enabled the construction of the new Devils Hole pupfish refuge. In contrast to most other federal land sales, where revenues revert to the U.S. Treasury, authors of SNPLMA, led by Nevada's congressional delegation, contrived to have the money stay in the state, with the bulk of the incoming funds—a total of over $3 billion by 2019—being spent on conservation projects.[93] The Devils Hole pupfish became the beneficiary of urban growth that ironically also offered a threat in the form of future regional groundwater decline.

On Being Lucky

All of these investments—in personnel, infrastructure, and management arrangements—were acts of redemption for a relationship between people and pupfish that had been seriously degraded over the preceding decades. In the years between 2006 and 2013, the Park Service, Fish and Wildlife Service, and Nevada Department of Wildlife reestablished and extended the bonds that had allowed the Devils Hole pupfish to survive the twentieth century. Instead of park policy or water law, however, these bonds required new agency relationships, procuring adequate funding, and making decisions with as much information as possible. It was not as exciting as marching toward the Supreme Court but every bit as important for the persistence of the species.

Consider how the agencies were able to stock the new Ash Meadows Fish Conservation Facility. Using protocols developed and tested in labs with captive hybrid pupfish—descendants of those accidentally created years earlier in the Point of Rocks Refuge—the agencies were able to agree on how to stock the refuge.[94] Instead of transferring adult pupfish, they made the choice to collect pupfish eggs, which they believed would be much less risky for the Devils Hole population. Managers planted "egg recovery mats" in Devils Hole during August and November 2013 and January 2014, recovering a total of sixty live pupfish eggs. Of the eggs collected, thirty-nine proved viable, thirty-three actually hatched, and twenty-nine were reared to adulthood in the aquarium tanks in the Fish Facility's laboratory.[95] Since then, the fish facility has provided an opportunity to learn about the pupfish, in addition to some "insurance" for managers. In 2018, managers began removing a species of diving beetle—also present in Devils Hole—from the fish facility tank that they had observed on camera predating pupfish eggs and larvae. Prior to attempts to control the beetle, managers struggled to harvest pupfish eggs in the artificial habitat. In their first attempt after removing some of the beetles, they collected forty eggs. Whether or not the diving beetle explains the decline of pupfish populations in Devils Hole is unknown, but it presents some tantalizing prospects for additional research.[96] Only continued monitoring and research will allow the Park Service, Fish and Wildlife Service, and Nevada Department of Wildlife to make a decision on whether it can be applied to Devils Hole itself.

The pupfish, for their part, do not appear to care about any of this noise. In April 2013, the Devils Hole pupfish population hit a new low with a scuba survey count tallying just thirty-five individuals. With the FCF yet to be stocked with pupfish, there were no refuge populations outside of Devils Hole; thirty-five pupfish constituted every known Devils Hole pupfish. Reflecting the strides that the agencies had made in improving their knowledge of the species, however, a *Las Vegas Review-Journal* reporter wrote, "What began decades ago as a fight to save the world's most isolated fish might soon end in one of the most well-documented extinctions ever."[97] By 2018 and 2019, however, the pupfish had the best year in recent memory: in fall 2018, the agencies counted 187 fish, and in spring 2019 they counted 136 fish. The numbers were the highest in at least fifteen years. The increases may be a result of specific actions taken by the agencies, including the supplemental feeding, but much remains unknown.[98] And the pupfish are not telling.

That the pupfish have persisted despite the extremely low population sizes recorded in the mid-2000s and in 2013 resurrects James Deacon's observation that the pupfish could "only be preserved if we learn enough about it to effect preservation—or failing that, only if we are lucky."[99] One way to read this statement is that without human institutions, investment, and caring we are unlikely to ensure the protection and survival of the pupfish, and perhaps all of nature. Only by remaining involved, curious, and committed, in other words, will we build relationships that will allow the persistence of the things we value and love. Indeed, this is the way that I have used his remark in this chapter, helping to frame the history of disinvestment in, and rebuilding of, the Park Service and its collaborators' relationship with the pupfish.

Yet there is a second, complementary, way to interpret "luck" with regard to the pupfish and other endangered species: not as a stand-in for the risky consequences of poor management practices of the let's-hope-for-the-best variety but as a word to communicate that pupfish are still wild animals. While human institutions have indeed enabled the pupfish's survival, the species is also beyond our control. The fish do not always behave as we expect, and they do not know or care about the Incident Command System, supplemental feeding regimes, or whether or not adequate public funding is thrown their way. They are, despite our best efforts to understand and manage them, partly unpredictable, wild.

As I read my way through the archival documents that undergird this book, I was treated to more than a century's worth of surprising records of how humans interacted with and shaped the future of the pupfish. But the first time the pupfish themselves surprised me personally, reminding me of their wildness, came far from a library reading room. In the spring of 2014, I toured the Hoover Dam refuge with Kevin Guadalupe, a Nevada Department of Wildlife biologist. We drove from his office in his NDOW pickup, winding through Las Vegas traffic toward Lake Mead and then passed through an automated security gate and eased down a narrow, paved road to the facil-

FIGURE 6.3. A Devils Hole pupfish found swimming in a creek below the Hoover Dam Refuge, April 2014. This individual was about twice the length of a typical pupfish in Devils Hole, suggesting it was the escaped offspring of hybrid pupfish reared at Hoover Dam Refuge in the early 2010s. The fish was returned to the habitat after this photograph was taken.

ity just downstream of the dam's face. The refuge facility had been empty of pupfish for years, but water continued to flow from a spring nestled in the rock wall above us into the concrete tank and then out toward the Colorado River below. Two empty blue fiberglass tanks sat nearby. They were added in 2006 as a temporary home for pupfish after the concrete tank became overrun with the invasive snail. Around the same time, the whole area had been covered with heavy mesh cargo netting to prevent predation from birds.

As we followed the small creek emerging from the refuge tank down toward the river, Guadalupe spotted something surprising: several pupfish (fig. 6.3). It may go without saying that Devils Hole pupfish are not supposed to be surviving, unmonitored and unstudied, in the pool of a tiny creek below an abandoned refuge that is reminiscent of a minor league batting cage. Guadalupe suspected that the pupfish were not pure Devils Hole pupfish but hybrids descended from the accidental introduction of Ash Meadows Amargosa pupfish to the Point of Rocks Refuge. In an effort to learn more about pupfish reproduction, some of these hybrids had been housed at Hoover Dam Refuge from about 2008 to 2011, before the effort was abandoned.[100] Regardless, outside of any official program, and basically unbeknownst to wildlife officials, pupfish continued to persist for years; "lucky," surprising, and wild.

CHAPTER 7

This Is Forever

For every plant and animal on the U.S. endangered species list, the federal government is supposed to publish a "recovery plan." The plans—not always carried out, as we have seen—provide consensus from experts on what actions should be taken to improve a species' prospects for survival, as well as what targets (typically measured in population size and range) must be achieved before it can be "delisted"—removed from the list and considered "recovered." The Devils Hole pupfish recovery plan, authored by James Deacon and others and then released by the Fish and Wildlife Service in 1980, is in large measure unremarkable in comparison with countless others.

But the Devils Hole plan stands out from its brethren in one crucial respect. Whereas these documents typically suggest what needs to be done for a species to be delisted, buried on page sixteen of the Devils Hole pupfish plan is a bombshell: "Even if pristine conditions were to be reestablished in Devils Hole," the Devils Hole pupfish could never be removed from the list itself, only downgraded to "threatened," a status that confers nearly identical legal protections.[1] The Devils Hole pupfish will always be threatened with extinction in the eyes of the federal government. This makes the pupfish the Hotel California of endangered species: it can be downgraded but it can never leave.

This judgment reflected not only the concern for the species' small population size and restricted habitat at the time of the recovery plan's writing—in the immediate wake of the Cappaert era in Ash Meadows—but also echoed scientists' longer-term fear of extinction at Devils Hole. Recall that when Robert Miller visited in June 1937, he declared, "It won't be long until they are extinct."[2] The recovery plan's statement about the pupfish's perpetually threatened status, even under "pristine" conditions, also in a way exemplifies the argument of this book: that over the course of the twentieth century the survival of the Devils Hole pupfish became deeply intertwined with some of the defining institutions and practices of the U.S. West, especially its science, water policy, and public land management. Eliminating one of these critical institutional threads—by delisting the species, for instance—could also begin to unravel the pupfish's survival. That the pupfish cannot be removed from the endangered species list is a bureaucratic way to admit that caring for the pupfish, and management of the environment around it, has become a forever job.

The managing agencies that oversee the pupfish have, it seems, finally embraced their long-term role at Devils Hole—something that has taken them decades to do. But simply acknowledging that monitoring, research, and management will continue indefinitely does not automatically tell managers how those increasingly detailed understandings of the Devils Hole ecosystem and the pupfish's genome should be interpreted. For the pupfish to continue to survive, we will likely have to continue to redefine our relationship with the species. The devil of the thing (if I may) is that the relationships we have already built with the pupfish are complicated and may only get more so. This is perhaps most true with two separate issues: climate change and the implications of new research on pupfish genetics.

Many of the key research papers on climate change and Devils Hole have been led by Mark Hausner, a hydrologist now at the University of Nevada's Desert Research Institute.[3] As part of his doctoral dissertation, Hausner investigated the effects of climate change on the Devils Hole pupfish, building a model to understand how water temperature on the shallow shelf changed under different future warming scenarios. He found that as the ambient air temperature in Ash Meadows rises—and it already has risen about 1°C over the last few decades—the window for successfully "recruiting" the next generation of pupfish will be shortened considerably. Increased air temperatures will make the water temperatures on the shallow shelf too high for pupfish eggs to hatch and juveniles to develop for longer stretches of the year. Moreover, the times of the year when water temperature on the shelf would be favorable for egg and larval development will not match the periods when adequate energy—pupfish food—is available in Devils Hole. The result of climate change in Devils Hole, Hausner found, will be a growing mismatch between periods of adequate temperatures and food availability.

This process is already underway. Hausner discovered that the window for pupfish recruitment has already shrunk from an estimated 74 days during the 1980s to 68 days between 2000 and 2010, perhaps contributing to the post-1995 population decline. And looking to the future, under a severe warming scenario resulting from "business as usual" carbon emissions, the window for pupfish recruitment could shrink to just 57 days a year by the end of this century—a reduction of some two weeks from the recent past.[4] For a species whose members live for only 10–14 months, this reduction in recruitment time could be catastrophic.

At Devils Hole, the global problem of increasing temperatures and the rising parts per million of carbon dioxide in the atmosphere interacts with the details of a tiny environment, producing this new, human-generated threat to the pupfish's future. But the species' survival of climate change will be conditioned by more than air temperature increases in the Mojave Desert, though of course this is critically important. The pupfish will survive—or not survive—these changes as a result of the deep, intertwined history it has already established with human institutions, and the choices we make in the future both in Nevada and globally.

As part of his work, Hausner ran his model of water temperatures on the shallow shelf not only imagining the water level at its current stage but also at its higher, pre-1968 level—before pumping by Cappaert's firm began to draw it down. He found something surprising. Even under extreme warming scenarios, if the water level in Devils Hole was restored to its pre-pumping level, the effects of climate change would be minimal: the recruitment window would actually be 75 days at the end of this century, slightly longer than during the 1980s when the pupfish population was high, and 18 days longer than his model forecasts for the year 2099 at the current water level.[5]

Hausner's alternative timeline, where pre-pumping water levels persist to the present and the pupfish have a longer season for recruiting the next generation, is in some ways conservative. A recent U.S. Geological Survey study found that if no groundwater pumping had ever occurred across the basins that influence Devils Hole, the water level in the pool would be several inches to a foot *higher* than prior to Cappaert's pumping, thanks to a period of increased precipitation in the mountains where the Devils Hole aquifer "recharges."[6] The pupfish is thus only threatened by global climate change if we ignore the decisions that led to the diminished water level in Devils Hole.

This insight returns us to the forever job, to the problem of how to enable the persistence of the Devils Hole pupfish. It would be difficult, though not impossible, to modify the *Cappaert* decision to force further restrictions in groundwater development. Even a total cessation of all pumping from the far-flung areas in southern Nevada that influence the Devils Hole water level, however, would not cause the water level to rise substantially. The same U.S. Geological Survey study mentioned above estimated that if all pumping halted in 2019, the water level in Devils Hole would only recover 0.1 ft. by 2100. If pumping continues as is, though, the water level could fall 1.5 feet from its current level by the end of the century.[7]

Such a reopening of the *Cappaert* decision to curtail future declines in the Devils Hole water level would also be politically explosive. New restrictions on groundwater pumping across southern Nevada for the Devils Hole pupfish and other species could provoke a backlash to federal power as it did in the 1970s with attending cries that the government puts pupfish ahead of people. Yet however difficult revising water use for pupfish and other aquatic species would be politically, it would at least be the starting point for an honest discussion of the threats facing the pupfish. Throwing up our hands and saying that the pupfish is endangered by the nebulous, global force of "climate change," a problem to which the entire fossil-fuel-consuming world contributes, is only part of the story. If the pupfish does not survive climate change, extinction will have as much to do with choices made in Nevada and its water system as with elevated concentrations of carbon dioxide in the atmosphere.

The second complex issue facing the future of Devils Hole pupfish management has to do with the implications of fresh arguments about the age of the species brought on by advances in genetic research. To appreciate this issue, it is important to review older assumptions of how and when pupfish colonized Devils Hole. For decades, the only work on the origins of the Devils Hole pupfish—when exactly ancestral pupfish arrived there—was conducted by Carl Hubbs and Robert Miller. As part of their efforts to understand the speciation of the Death Valley pupfishes, they began piecing together a history of interconnected Pleistocene lakes that gradually receded and left pupfish stranded in places such as the Owens Valley, Salt Creek, and, of course, Devils Hole.[8] A general version of this interpretation made its way into President Truman's 1952 proclamation adding Devils Hole to Death Valley National Monument. The proclamation imparted value to the pupfish due to it having "evolved only after the gradual drying up of the Death Valley Lake System isolated this fish population from the original ancestral stock that in Pleistocene times was common to the entire region."[9]

Not only was the pupfish unique but it was also old, evidence of a wetter, "pluvial" period in the Death Valley region's history. As for an actual date at which the pupfish became isolated in Devils Hole, Miller only sometimes offered a specific number in publication, suggesting 100,000 years ago in one article and between 10,000 and 20,000 years ago in another.[10]

Over the decades, geological studies refined the window for potential introduction of the pupfish to Devils Hole, but the new knowledge only made the story more puzzling. Geologists learned that the roof over Devils Hole collapsed (and the pool thus opened to the sky) about 60,000 years ago, meaning that the pupfish must have arrived sometime after that date.[11] Analyses of Pleistocene waters in the Death Valley region, however, revealed that lakes did not reach anywhere near as high as Devils Hole after the cavern opened to air. Additionally, scientists have found no evidence that the water level in Devils Hole was ever high enough to overflow its crevasse and spill downhill toward the rest of Ash Meadows (and thus allow fish to swim upstream from a body of water below). The picture was muddied enough that USGS's Alan Riggs and UNLV's James Deacon declared that while scientists could explain the multimillion-year journey of ancestral pupfish from the East Coast and Caribbean into isolated, inland river basins of the West and all the way to Ash Meadows, "there is no obvious way for [pupfish] to have jumped the apparently dry last kilometer to get into Devils Hole!"[12]

This is where the question of the pupfish's origin stood in 2013 when I read Riggs and Deacon's quip and began researching the species' modern history. And while intriguing, it seemed to me to be of little relevance to my work or to the future of the pupfish, regardless of how the puzzle was solved. But starting in 2014, a string of scientific papers using advanced genetic techniques began to propose answers to this last kilometer problem, providing fresh interpretations with far-reaching implications for the pupfish's future.

Strictly speaking, the papers attempted to calculate the likely "age" of the Devils Hole pupfish based on how far it had mutated away from its close genetic relatives, although they implicitly addressed the question of when and how the fish became isolated. (If you know the rate at which mutations occur, then you can use comparisons between the genomes to estimate the time since the species became geographically isolated from one another.[13]) The first paper, by J. Michael Reed and Craig Stockwell, suggested that the Devils Hole pupfish diverged from its close relative, the Ash Meadows Amargosa pupfish, between 217 and 2,530 years ago—much more recently than any previous estimate. Two years later, another study, led by Christopher Martin, suggested the Devils Hole pupfish was between just 105 years and 830 years old, with 255 years their best estimate.[14] Rather than an ancient holdover from the Pleistocene, the Devils Hole pupfish may be younger than the Code of Hammurabi, Magna Carta, and, perhaps, the Declaration of Independence.

To some, the possibility that the pupfish is a recent arrival to Devils Hole has made the species' future more doubtful. Conservation biology suggests that genetic problems facing small, isolated populations mean they are unlikely to survive very long. Martin and his coauthors concluded their paper by writing that while the post-1995 pupfish population decline was not likely the result of a high "genetic load" of problematic mutations, "theory predicts that extinction due to increasing genetic load is highly likely for this small population over timescales of 100 years or less."[15]

Just months after Christopher Martin's paper estimated the pupfish's age at just 255 years, however, a third paper, by a University of California, Davis, research team led by İsmail Sağlam, used slightly different techniques and assumptions to suggest that pupfish may have colonized Devils Hole in the same (but undetermined) event that caused the cavern to collapse and exposed it to air 60,000 years ago.[16] Sağlam's analysis confirmed, in other words, the assumptions of Robert Miller, which were captured in Truman's presidential proclamation: the pupfish are descended from the Pleistocene era. In this interpretation, the pupfish would have already passed through a "genetic bottleneck," and having purged damaging mutations from its genome could thus be a confirmed outlier for conservation biology: a remarkably small population that has persisted for a very long time. In 2017 and 2018, Martin and Sağlam traded barbs and critiques of each other's work in two different scientific journals, with both scientists sticking to their respective guns and leaving the timing of the origins of the pupfish somewhere between 60,000 and a couple hundred years ago.[17]

The notion that the pupfish is a recent arrival to Devils Hole is provocative. If demonstrated more definitively, it would undermine part of the rationale used in the Truman proclamation, while simultaneously deepening the mystery and wonder of how the pupfish have colonized and survived in this extreme habitat. None of the recent papers on the age of the pupfish had any real mechanism baked into their analyses for explaining these results—i.e., *how* pupfish got into Devils Hole. Christopher Martin has suggested it is possible that an extreme flooding event enabled pupfish to reach the hole or that pupfish eggs were inadvertently deposited in Devils Hole after being stuck to bird legs.[18] (This second suggestion, while entirely plausible, always reminds me of *Monty Python and the Holy Grail*, where castle guards debate how coconuts could have arrived in England: perhaps carried by swallows?) Reed and Stockwell, the authors of the first recent-origins paper, referenced another hypothesis: "Native Americans may have intentionally or unintentionally introduced pupfish to Devils Hole."[19]

I am very intrigued by this possibility. After all, the bulk of this book describes how the Devils Hole pupfish has persisted, not by being isolated but by becoming entangled in the modern institutions of the West. What if, then, humans actually played a role in its speciation? The notion suggests that we have the power not only to destroy—amply demonstrated—but also, perhaps, to create. It would also imply that our duty to protect this species has to do with our role in its origins. The pupfish might always have survived through its relationship with us, not only since the late nineteenth century as this book has explored.[20]

But even if the pupfish was a recent introduction, and humans played a role in that arrival, it still would not tell managers what to do next. At the 2018 Desert Fishes Council meeting, Christopher Martin, one of the authors whose analysis suggests a recent origin of the pupfish, gave a presentation on his work. Concluding, Martin said that one way to manage the Devils Hole pupfish in the future would be to restore the genome of the pupfish population in the Fish Conservation Facility by using the "CRISPR" gene-editing technique to insert older genes pulled from museum specimens collected by Robert Miller in the 1930s.[21] This would have the effect of reintroducing some genetic diversity, perhaps extending the life of the species.

Christopher Martin's proposal was not the first time a geneticist named Martin proposed something like this. Several years earlier another, unrelated pupfish geneticist named, confusingly, Andrew Martin, suggested a much lower-tech option for genetic restoration. Played out in a *Wired* magazine cover story, Andrew Martin told a reporter that a way to rescue the Devils Hole pupfish from extinction would be to hybridize it with Ash Meadows Amargosa pupfish, its close relative, by dropping one or two individuals into Devils Hole. For the Park Service, this was a nonstarter: "It wouldn't be the same species," Kevin Wilson told the reporter.[22] (Under the Endangered Species Act, hybridization has been frowned on except in a few cases, such as the Florida panther.[23])

There is no easy answer for whether or not to take up the recommendations of Christopher Martin or Andrew Martin. Altering the pupfish genome, whether by CRISPR or hybridization, may seem extreme, but it would represent only the latest—not the first—step in our relationship with the species. Doing so would, however, mean accepting the recent-origins interpretation of Christopher Martin's study. If the pupfish have been in Devils Hole for 60,000 years, then genetic restoration would not seem necessary. So far, the jury is still out.[24] Ultimately, even as we have become deeply enmeshed with the survival of the Devils Hole pupfish, prescriptions for how to act in the future are not always obvious, something both climate change and pupfish genetics vividly illustrate.

It can seem hubristic to say that we will always be responsible for the pupfish—as if our will alone can control its fate. People who care about the Devils Hole pupfish sometimes express this feeling. While deeply sympathetic to efforts to protect the Devils Hole pupfish from extinction, ecologist and science writer Christopher Norment worries that "we simply are too involved in the lives of those little fish" and expresses sadness that "we believe we can master the natural world enough to control the fate of the species."[25] I respect Norment and enjoy his thoughtful and deeply personal writing on desert species and natural history, but I disagree with him here. All the memos, letters, and reports written by those who have been closest to the pupfish—the evidence that I have pored over for years and now forms much of this book's narrative—do not seem to me to be written by masters. Instead, if I could characterize all those voices, I would say that most have labored mightily to grasp a sliver of truth from a system so complex, despite its tiny size, that they doubted their ability to capture much more. The forever job, as I see it, is not about mastery but humility—even if it does involve a $4.5 million quasi-replica of Devils Hole.

A few years ago, a much more basic (and base) critique of pupfish management found its way into the *Las Vegas Review-Journal.* Complaining about the amount of money spent on the Ash Meadows Fish Conservation Facility, the newspaper's editorial board asked, "Couldn't scarce government resources be put to more productive environmental uses while Mother Nature sorts out the future of the Devils Hole pupfish?"[26] I say no. We have become so intertwined with the pupfish's recent history that even the decision to leave it to "Mother Nature"—if it were even institutionally possible given the requirements of the Endangered Species Act and other decisions—is not so much wrong as it is all but meaningless. We have defined, shaped, threatened, and enabled the pupfish's survival at so many key junctures that walking away now would itself be a management choice, and any future harm would not be nature running its course but a societal decision.

The other part of the *Review-Journal*'s concern lay, somewhat disingenuously, with funding; the title of the editorial was "Do We Have Unlimited Resources to Save the Devils Hole Pupfish?" Aside from the fact that dozens of threatened and endangered species receive much more funding than the Devils Hole pupfish, this is an old critique in the story of the pupfish, dating back to the Cappaert-era in Ash Meadows.[27] As described in the introduction to this book, back in 1970 a senior Bureau of Sport Fisheries and Wildlife official reviewed plans to protect the Devils Hole pupfish and then wrote, "Such a crash program may well save the pupfish...but what of the other 80+ species also on the list?"[28] The editorial writers at the newspaper were only the latest to consider whether we can expect the same commitment to other environmental issues as we have devoted to the pupfish.

The Devils Hole pupfish is exceptional in the way that it became intertwined with nineteenth- and twentieth-century human history. As I have highlighted, this pupfish survived when other species have not: the Tecopa pupfish, the Las Vegas dace, and the Ash Meadows poolfish—among many others—did not survive the twentieth century, and many more will face the same fate as we accelerate our way into the "sixth extinction." The pupfish got incredibly lucky to be classified, protected in a presidential proclamation, recognized by the state water system, managed by public agencies, and cared for by generations of public employees.

But just because the pupfish was lucky, does not mean we should not expect as much of ourselves when it comes to other species and landscapes. We do not need to view the pupfish's history dogmatically as the only model for protecting species and places. If we strip this story to its barest parts, what we have are several generations of people—from T. S. Palmer, to Robert Miller and Carl Hubbs, to Phil Pister, and all the way to J. T. Reynolds and Kevin Wilson—who developed and repeatedly reworked our relationship with the pupfish to enable its survival. To me, this echoes an idea from Wes Jackson, founder of the Land Institute, who has commented that one of the ways to ensure we take care of agricultural and forested land is to increase its "eyes-to-acres ratio."[29] In other words, if we seek to care for a place, restore its soil, and ensure a healthy ecosystem, it is important to have plenty of people paying attention to it. While Jackson is focused on productive landscapes, the same might be true for other kinds of landscapes as well. The more people who are concerned with a place and its biodiversity, the greater the likelihood that it will be cared for well. Perhaps, then, the thing that has made the pupfish so lucky, or unique, is that it has had many people over the years focused on a very small area—a lot of eyes per square foot.

Yet the relationships between pupfish and people that have allowed the species to persist through modern history are no guarantee of its future survival. And although I have now written a book about the species' persistence, I know that the pupfish is not bulletproof. Even if we can remain committed "forever," the pupfish may become extinct as the dark side of our tangled history catches up to it, whether due to our carbon emissions and groundwater pumping, our imperfect knowledge of its genetics and ecology, or something as yet unforeseen. Should extinction be the end of the story for *Cyprinodon diabolis*—tomorrow, in a hundred years, or in ten thousand years—I hope one of the questions asked by some journalist or curious Nevadan is not only what brought about this sad denouement, but what allowed the Devils Hole pupfish to live, to survive. I think they will find in answering this question—at least for the roughly 130 years considered here—a story of a flawed, unequal symbiosis, where people made room for the pupfish in their lives, practices, and institutions.

I first visited Devils Hole on a slate-gray February afternoon in 2014, adding my own set of eyes to a long list of others who have considered the pupfish. I had driven across most of the country in a borrowed car, spending a chilly night in a Colorado hotel room and shivering through an even colder one in my tent in northern Utah. Then, after a long week of reading through documents in a Las Vegas library, I left the sprawl and the highways and bounced over several miles of dirt roads to see Devils Hole for myself. I walked the last 200 yards to the fence installed by Death Valley National Park, finally peering down to the tepid pool (fig. 7.1). It was a bit of a letdown.

FIGURE 7.1. Workers install the first visitor viewing platform and fence around Devils Hole, 1965. Resource Management Collection, Death Valley National Park Library and Archives, National Park Service.

Essentially, on that first visit I did not know the story of how the pupfish had become intertwined in our modern world. As I have gone back to Devils Hole over the years and stood in that same place on the visitor platform, I have had the sensation that if I could just squint a little more, I would see the strands of human history that have enabled the pupfish to survive so far. Not quite, but almost.

List of Abbreviations

BLM	Bureau of Land Management
BSFW	Bureau of Sport Fisheries and Wildlife
ESA	Endangered Species Act of 1973
FCF	Ash Meadows Fish Conservation Facility
FWS	Fish and Wildlife Service
ICS	Incident Command System
NDFG	Nevada Department of Fish and Game
NDOW	Nevada Department of Wildlife
NPS	National Park Service
UC-BERKELEY	University of California, Berkeley
UCSD	University of California, San Diego
UNLV	University of Nevada, Las Vegas
USDA	U.S. Department of Agriculture
USGS	U.S. Geological Survey

Notes

Introduction

1. A quick Google search reveals that the name "Devils Hole" has been given to sites across the United States. Sadly, I have turned up precious little on when, why, and who first applied the name to the site that is at the heart of this book. The earliest instance of its written usage that I have discovered comes from the field notes of Theodore Sherman Palmer, who visited Devils Hole in 1891. But it is clear from the context that settlers in Ash Meadows were already using this name when he arrived. See Theodore Sherman Palmer, diary, March 20, 1891, HM50827, Theodore S. Palmer papers, Huntington Library, San Marino, California (hereafter Palmer papers, Huntington).

2. W. W. Dudley Jr. and J. D. Larson, *Effect of Irrigation Pumping on Desert Pupfish Habitats in Ash Meadows, Nye County, Nevada*, Geological Survey Professional Paper 927 (Washington, DC: GPO, 1976), 5–12.

3. James Deacon, interview by Kevin C. Brown, April 3, 2014, Henderson, NV, transcript, author's possession, 17.

4. Alan C. Riggs and James E. Deacon, "Connectivity in Desert Aquatic Ecosystems: The Devils Hole Story," in *Conference Proceedings, Spring-Fed Wetlands: Important Scientific and Cultural Resources of the Intermountain Region, 7–9 May 2002*, edited by D. W. Sada and S. E. Sharpe, DRI Report 41201 (Reno: Desert Research Institute, 2004), 4. On the details of the trespassing divers, and the failed rescue and recovery effort, see NPS correspondence in box 607, P11, RG 79, NARA-College Park.

5. Dudley and Larson, *Effect of Irrigation Pumping*, 12. The size of the Ash Meadows section of the Amargosa Desert is poorly defined. Most of the springs (and species) occur across a smaller area than the 50,000 acres cited above—mostly within about 20,000 acres that are now part of Ash Meadows National Wildlife Refuge. However, the U.S. Fish and Wildlife Service uses the larger figure to capture the broader ecosystem. See Don W. Sada, *Recovery Plan for the Endangered and Threatened Species of Ash Meadows, Nevada* (Portland, OR: U.S. Fish and Wildlife Service, 1990), 3, 37–38.

6. Kathie Taylor, *Ash Meadows: Where the Desert Springs to Life* (Death Valley, CA: Death Valley Natural History Association, 2015), 39; and Bruce A. Stein, Lynn S. Kutner, and Jonathan S. Adams, *Precious Heritage: The Status of Biodiversity in the United States* (New York: Oxford University Press, 2000), 198. The Galápagos Islands have their own complex history of the relationships between humans and biodiversity wonderfully captured in Elizabeth Hennessy, *On the Backs of Tortoises: Darwin, the Galapagos, and the Fate of an Evolutionary Eden* (New Haven: Yale University Press, 2019).

7. Kevin P. Wilson and Dean W. Blinn, "Food Web Structure, Energetics, and Importance of Allochthonous Carbon in a Desert Cavernous Limnocrene: Devils Hole, Nevada," *Western North American Naturalist* 67, no. 2 (2007):185–198.

8. Alan C. Riggs and James E. Deacon, "Connectivity in Desert Aquatic Ecosystems: The Devils Hole Story," in Sada and Sharpe, *Conference Proceedings, 2002, Spring-Fed Wetlands*, 16–19.

9. Louis Nusbaumer, in *Escape from Death Valley: As Told by William Lewis Manly and Other '49ers*, edited by Leroy Johnson and Jean Johnson (Reno: University of Nevada Press, 1987), 160.

10. James E. Deacon and Cynthia Deacon Williams, "Ash Meadows and the Legacy of the Devils Hole Pupfish," in *Battle Against Extinction: Native Fish Management in the American West*, edited by W.L. Minckley and James E. Deacon (Tucson: University of Arizona Press, 1991), 69.

11. U.S. Department of Energy, *United States Nuclear Tests: July 1945 through September 1992*, DOE/NV-209-REV 15 (2000), xv; Energy Research and Development Administration, *Nevada Test Site, Nye County, Nevada: Final Environmental Impact Statement*, September 1977, 4–28; and Howard H. Chapman to Regional Director, FWS, Portland, OR, February 25, 1976, series 008, pupfish activities, resource management collection, National Park Service, Death Valley National Park archives, Cow Creek, California (hereafter pupfish activities, NPS-Cow Creek).

12. On the notion that an "endangerment sensibility" pervades a number of disparate fields beyond just endangered species science and management, see Fernando Vidal and Nélia Dias, eds., *Endangerment, Biodiversity, and Culture* (New York: Routledge, 2016), 2. On the cultural significance of endangerment, extinction, and reintroduction, see Ursula K. Heise, *Imagining Extinction: The Cultural Meaning of Endangered Species* (Chicago: University of Chicago Press, 2016); and Dolly Jørgensen, *Recovering Lost Species in the Modern Age: Histories of Longing and Belonging* (Cambridge, MA: MIT Press, 2019).

13. On recent work on extinction and the idea of a sixth mass extinction event, see Anthony D. Barnosky et al., "Has the Earth's Sixth Mass Extinction Already Arrived?" *Nature* 471 (March 3, 2011):51–57; Gerardo Ceballos, Paul R. Ehrlich, and Rodolfo Dirzo, "Biological Annihilation via the Ongoing Sixth Mass Extinction Signaled by Vertebrate Population Losses and Declines," *PNAS* 114, no. 30 (July 2017):E6089–E6096, doi:10.1073/pnas.1704949114; and Malcolm L. McCallum, "Vertebrate Biodiversity Losses Point to a Sixth Mass Extinction," *Biodiversity and Conservation* 24, no. 10 (2015):2497–2519, doi:10.1007/s10531-015-0940-6. For earlier discussions of a "sixth extinction," see Edward O. Wilson, *The Diversity of Life* (Cambridge, MA: Harvard University Press, 1992); and Richard Leakey and Roger Lewin, *The Sixth Extinction: Patterns of Life and the Future of Humankind* (New York: Doubleday, 1995). Also see Elizabeth Kolbert's excellent journalistic review of biodiversity loss and the sixth extinction, *The Sixth Extinction: An Unnatural History* (New York: Henry Holt, 2014).

14. Intergovernmental Science-Policy Platform on Biodiversity and Ecosystem Services (IPBES), *Summary for Policymakers of the Global Assessment Report on Biodiversity and Ecosystem Services* (2019), 14n4, https://zenodo.org/record/3553579.

15. On the history and application of population viability analysis, see Fred Van Dyke, *Conservation Biology: Foundations, Concepts, Applications*, 2nd ed. (Springer: 2008), 224–242.

16. Steven R. Beissinger, "Digging the Pupfish Out of Its Hole: Risk Analysis to Guide Harvest of Devils Hole Pupfish for Captive Breeding," *PeerJ Life and Environment* 2:e549 (2014), doi:10.7717/peerj.549.

17. Robert R. Miller to Carl L. Hubbs, June 22, 1937, folder 11, box 230, Carl L. Hubbs Papers (81–18), Scripps Institution of Oceanography archives, University of California, San Diego (hereafter, Hubbs papers, UCSD).

18. I have been inspired by Kate Brown's (no relation) "embodied approach" to writing

history. See her *Dispatches from Dystopia: Histories of Places Not Yet Forgotten* (Chicago: University of Chicago Press, 2015), esp. 1–18.

19. For a reflection on case-study selection, see Jason Seawright and John Gerring, "Case Selection Techniques in Case Study Research: A Menu of Qualitative and Quantitative Options," *Political Research Quarterly* 61, no. 2 (June 2008):294–308.

20. J.P. Linduska, handwritten note, n.d., stapled to front of James McBroom to Director, BSFW, August 24, 1970, FWS-Las Vegas.

Chapter 1. What's in a Name

1. *Pumpkin seed*: C. Hart Merriam, "North American Fauna No. 6 1890s," box 1, folder 1, Clinton Hart Merriam papers (MVZA.MSS.0281), Museum of Vertebrate Zoology, University of California, Berkeley, p. 234, unpublished (hereafter, Merriam, "North American Fauna No. 6 1890s"). *Pursy minnow*: David Starr Jordan and Barton Warren Evermann, *The Fishes of North and Middle America: A Descriptive Catalogue of the Species of Fish-Like Vertebrates Found in the Waters of North American, North of the Isthmus of Panama*, part 1, Bulletin of the United States National Museum, no. 47 (Washington, D.C.: GPO, 1896), 670. *Pupfish*: coined by Carl Hubbs, first found in manuscript from 1945, first used in published paper in 1950. See Robert R. Miller to Carl Hubbs, October 8, 1971, folder 18, box 55, Hubbs papers, UCSD; Robert R. Miller, "List of the Fishes of Nevada," October 22, 1945, folder "Nevada Checklists and Keys," box "Regional Index—Mexico-Nevada," Robert R. Miller papers, Fishes Division, University of Michigan Museum of Zoology (hereafter, Miller papers, UMMZ); and Leo Shapovalov and William A. Dill, "A Check List of the Fresh-Water and Anadromous Fishes of California," *California Fish and Game* 36, no. 4 (1950):387.

2. On the necessity, invisibility, and functions of classification across many fields and disciplines, see Michel Foucault, *The Order of Things: An Archaeology of the Human Sciences* (New York: Vintage, 1973); and Geoffrey C. Bowker and Susan Lee Star, *Sorting Things Out: Classification and Its Consequences* (Cambridge, MA: MIT Press, 1999). On classification in the history of science, see Harriet Ritvo, *The Platypus and the Mermaid and Other Figments of the Classifying Imagination* (Cambridge, MA: Harvard University Press, 1997).

3. "Museum Support Center: Media Fact Sheet," https://www.si.edu/newsdesk/factsheets/museum-support-center. Statistics on fishes collection from "Summary Statistics, Division of Fishes," printed handout from Museum Support Center, in author's possession.

4. Peter S. Alagona, "Species Complex: Classification and Conservation in American Environmental History," *Isis* 107, no. 4 (2016):761.

5. Theodore Sherman Palmer, diary, March 20, 1891, Palmer papers, Huntington. Palmer's fish collection is currently stored in alcohol at the Smithsonian, but it is not certain how he initially preserved the specimens. Into the 1880s, it was common to preserve fishes immediately in alcohol, but by the early twentieth century ichthyologists typically "fixed" specimens for a period of time in formalin before transferring the specimens to alcohol. See Tarleton H. Bean, *Directions for Collecting and Preserving Fish* (Washington, D.C.: GPO, 1881); and David Starr Jordan, *A Guide to the Study of Fishes*, vol. 1 (New York: Henry Holt, 1905), 432–433.

6. Palmer continued to disappoint me with his apparent lack of interest in Devils Hole. I learned that near the end of his life he produced a study of the origins of Death Valley area place-names. Devils Hole did not make it into his book. See T.S. Palmer, ed., *Place Names of the Death Valley Region in California and Nevada* (1948), https://hdl.handle.net/2027/uc1.31822035077684.

7. Keir B. Sterling, "Builders of the U.S. Biological Survey, 1885–1930," *Journal of Forest History* 33, no. 4 (1989):180–187. See also Jenks Cameron, *The Bureau of Biological Survey: Its History, Activities, and Organization*, Institute for Government Research, Service Monographs of the United States Government 54 (Baltimore: Johns Hopkins Press, 1929).

8. Keir B. Sterling, *Last of the Naturalists: The Career of C. Hart Merriam* (New York: Arno Press, 1974).

9. C. Hart Merriam, *Results of a Biological Survey of the San Francisco Mountain Region and Desert of the Little Colorado, Arizona*, North American Fauna No. 3 (Washington, D.C.: GPO, 1890).

10. Jane Maienschein, "Pattern and Process in Early Studies of Arizona's San Francisco Peaks," *BioScience* 44, no. 7 (1994):480; and Roderick P. Neumann, "Life Zones: The Rise and Decline of a Theory of the Geographic Distribution of Species," in *Spatializing the History of Ecology: Sites, Journeys, Mappings*, edited by Raf de Bont and Jens Lachmund (New York: Routledge, 2017), 37–55. Merriam's work echoed, for example, Alexander von Humboldt's concept of *Naturgemälde* and his study of plant distribution around the peak of Chimborazo in modern-day Ecuador. See Andrea Wulf, *The Invention of Nature: Alexander von Humboldt's New World* (New York: Vintage, 2015), 98–107, esp. 101–103.

11. Merriam, "North American Fauna No. 6 1890s," 126. The first part of the Death Valley Expedition report, known as North American Fauna No. 6, was never published.

12. On nineteenth-century government surveys in the West, see Wallace Stegner, *Beyond the Hundredth Meridian: John Wesley Powell and the Second Opening of the West* (Boston: Houghton Mifflin, 1954; repr. New York: Penguin, 1992); Robert E. Kohler, *All Creatures: Naturalists, Collectors, and Biodiversity, 1850–1950* (Princeton, NJ: Princeton University Press, 2006); Jeremy Vetter, *Field Life: Science in the American West during the Railroad Era* (Pittsburgh, PA: University of Pittsburgh Press, 2016); William H. Goetzmann, *Exploration and Empire: The Explorer and the Scientist in the Winning of the American West* (New York: Knopf, 1966); and Michael L. Smith, *Pacific Visions: California Scientists and the Environment, 1850–1915* (New Haven, CT: Yale University Press, 1987).

13. On structure and operation of the Death Valley Expedition, see Merriam, "North American Fauna No. 6 1890s."

14. Kohler, *All Creatures*, 184.

15. Hal K. Rothman and Char Miller, *Death Valley National Park: A History* (Reno: University of Nevada Press, 2013), 8–10; Heidi Roberts and Richard V.N. Ahlstrom, eds., *A Prehistoric Context for Southern Nevada*, Archaeological Report No. 011-05, (Las Vegas: HRA Inc., 2012); and Douglas Deur and Deborah Confer, *People of the Snowy Mountain, People of the River: A Multi-Agency Ethnographic Overview and Compendium Relating to Tribes Associated with Clark County, Nevada*, Pacific West Region, Social Science Series 2012-01 (National Park Service, 2012).

16. D.A. Lyle, "Appendix B: Report of Second Lieutenant D.A. Lyle, Second United States Artillery," in A.A. Humphreys, *Preliminary Report Concerning Explorations and Surveys Principally in Nevada and Arizona… From Brigadier General A.A. Humphreys, Chief of Engineers, Conducted under the Immediate Direction of 1st Lieut. George M. Wheeler, Corps of Engineers, 1871* (Washington, D.C.: GPO, 1872), 89. On Southern Paiute and Western Shoshone in the nineteenth century, see David Hurst Thomas, Lorann S.A. Pendleton, and Stephen C. Cappannari, "Western Shoshone," in *Handbook of North American Indians*, vol. 11, *Great Basin* (Washington, D.C.: Smithsonian, 1986), 262–283; Isabel T. Kelly and Catherine S. Fowler, "Southern Paiute," in *Handbook of North American Indians*, vol. 11, *Great Basin*, 368–397; and Martha C. Knack,

Boundaries Between: The Southern Paiutes, 1775–1995 (Lincoln: University of Nebraska Press, 2001).

17. Frederick Vernon Coville, "The Panamint Indians of California," *American Anthropologist* 5, no. 4 (October 1892):358.

18. Theodore Sherman Palmer, diary, March 5, 1891, Palmer papers, Huntington. On the material and cultural relationships between the Southern Paiute and springs, see Catherine S. Fowler, "What's in a Name? Some Southern Paiute Names for Mojave Desert Springs as Keys to Environmental Perception," in Sada and Sharpe, *Conference Proceedings, Spring-Fed Wetlands.*

19. Theodore Sherman Palmer, diary, March 20, 1891, Palmer papers, Huntington.

20. T. S. Palmer photograph collection, box 1, item 23, photCL416, Huntington. Other attitudes toward Native Americans evinced by the Death Valley Expedition also reflect broader cultural attitudes of the era, viewing Native Americans as either a threat toward wildlife (because of subsistence hunting practices) or as nearly extinct cultures. For these views, see *The Death Valley Expedition: A Biological Survey of Parts of California, Nevada, Arizona, and Utah*, part 2, North American Fauna No. 7, Division of Ornithology and Mammalogy, U.S. Department of Agriculture (Washington: GPO, 1893), 148 (hereafter USDA, *The Death Valley Expedition*); and Frederick Vernon Coville, "The Panamint Indians of California," esp. 351–352.

21. Rothman and Miller, *Death Valley National Park*, 16–27; and Richard E. Lingenfelter, *Death Valley and the Amargosa: A Land of Illusion* (Berkeley: University of California Press, 1986).

22. For descriptions of how the Death Valley Expedition interacted with and relied on the industrializing economy around the region, see Merriam, "North American Fauna No. 6 1890s"; and Vernon O. Bailey, "Into Death Valley 50 Years Ago," *Westways* (December 1940):8–11.

23. Theodore Sherman Palmer, diary, February 1, 1891, Palmer papers, Huntington.

24. "In the Valley of Death: The Work of the Expedition in That Region Completed," *The Examiner* (San Francisco), June 7, 1891, 11.

25. Merriam, "North American Fauna No. 6 1890s," 129.

26. Calculation of number of total fish specimens collected by the Death Valley Expedition based on consulting Gilbert's report and a search of ichthyology database fishnet2.net. I manually filtered the results of a search for all specimens with a collection date of 1891 held at the Smithsonian National Museum of Natural History.

27. Hennessy, *On the Backs of Tortoises*, 76. See also Elizabeth Hennessy, "Saving Species: The Co-Evolution of Tortoise Taxonomy and Conservation in the Galápagos Islands," *Environmental History* 25 (2020):263–286.

28. On laboratory and field biology in the early twentieth century, see Robert E. Kohler, *Landscapes and Labscapes: Exploring the Lab-Field Border in Biology* (Chicago: University of Chicago Press, 2002).

29. Merriam, "North American Fauna No. 6 1890s," 234.

30. On Gilbert's career, see J. Richard Dunn, "Charles Henry Gilbert (1859–1928): Pioneer Ichthyologist of the American West," in Theodore W. Pietsch and William D. Anderson Jr., eds., *Collection Building in Ichthyology and Herpetology* (American Society of Ichthyologists and Herpetologists, 1997), 265–278.

31. David S. Jordan and Charles H. Gilbert, *Synopsis of the Fishes of North America*, Bulletin of the United States National Museum No. 16 (Washington, D.C.: GPO, 1882).

32. Charles Henry Gilbert and Norman Bishop Scofield, "Notes on the Fishes from the Colorado Basin in Arizona," *Proceedings of the United States National Museum* 20 (Washington, GPO, 1898), 487.

33. David Starr Jordan, "Charles Henry Gilbert," *Science* 67, no. 1748 (June 29, 1928):645.

34. J. Percy Baumberger, "A History of Biology at Stanford University," *Bios* 25, no. 3 (1954): 123–147.

35. Michael Espinosa and Benjamin Zaidel, "Stanford to Rename Spaces Honoring David Starr Jordan, Founding President and Noted Eugenicist," *Stanford Daily*, October 7, 2020, https://www.stanforddaily.com/2020/10/07/stanford-to-rename-spaces-honoring-david-starr-jordan-founding-president-and-noted-eugenicist/.

36. Carl L. Hubbs, "History of Ichthyology in the United States after 1850," *Copeia* 1964, no. 1 (March 26, 1964):48. On natural history museums, see Stephen T. Asma, *Stuffed Animals and Pickled Heads: The Culture and Evolution of Natural History Museums* (New York: Oxford University Press, 2001).

37. Charles H. Gilbert, "Report on the Fishes of the Death Valley Expedition Collected in Southern California and Nevada in 1891, with Descriptions of New Species," in USDA, *The Death Valley Expedition*, 229–234.

38. Jordan and Evermann, *The Fishes of North and Middle America*, 670.

39. After the Mexican-American War, the U.S. government conducted a survey to confirm the new boundary with Mexico. On its return, a Frenchman, physician, and later a Confederate sympathizer named Charles Girard identified the fishes collected by members of the expedition. After examining eight specimens from the Rio San Pedro in southern Arizona, Girard identified a new species—*Cyprinodon macularius*. See Charles Girard, "Ichthyology of the Boundary," in *United States and Mexican Boundary Survey, under the Order of Lieut. Col. W.H. Emory, Major First Cavalry, and United States Commissioner*, vol. 2, part 6 (Washington, D.C.: Cornelius Wendell, 1859), 68, http://hdl.handle.net/2027/miun.afk4546.0002.006.

40. Charles H. Gilbert, "Report on the Fishes of the Death Valley Expedition," 232.

41. Ibid.

42. Ibid., 233.

43. Carolus Linnaeus [Karl von Linné], *Systema Naturae*, eds. M. S. J. Engel-Ledeboer and H. Engel (1735; repr. Nieuwkoop, NL: B. de Graff, 1964); Wilfrid Blunt, *Linnaeus: The Compleat Naturalist*, rev. ed. (Princeton, NJ: Princeton University Press, 2002); and Lisbet Koerner, *Linnaeus: Nature and Nation* (Cambridge, MA: Harvard University Press, 1999). On species classification, see also Pierre Pellegrin, *Aristotle's Classification of Animals: Biology and the Conceptual Unity of the Aristotelian Corpus*, trans. Anthony Preus (Berkeley: University of California Press, 1986); John S. Wilkins, *Species: A History of the Idea* (Berkeley: University of California Press, 2009); and Carol Kaesuk Yoon, *Naming Nature: The Clash between Instinct and Science* (New York: Norton, 2009).

44. For an overview of the history of evolution and the integration of Darwin's natural selection theory with population genetics in the early twentieth century (the "modern synthesis"), see Peter J. Bowler, *Evolution: A History of the Idea*, 25th anniversary edition (Berkeley: University of California Press, 2009).

45. See, for example, Jordan, *Guide to the Study of Fishes*, 367–368.

46. Elliot Coues, *Key to North American Birds* (Boston: Estes and Lauriat, 1884), 79–80. Coues reprinted this point in various later volumes. This specific paragraph was first printed in a letter to *The Auk*: Coues, letter to the editor, "Trinomials are Necessary," *The Auk* 1, no. 2 (April 1884):197–198.

47. Jordan quoted Coues verbatim in his *Guide to the Study of Fishes*, 379.

48. David Starr Jordan, "The Origin of Species through Isolation," *Science* 22, no. 566 (Nov. 3, 1905):561–562.

49. Jordan and Evermann, *The Fishes of North and Middle America*, v. In practice, it is very difficult to build perfect lineages of evolutionary relatedness. As a 1962 ichthyology textbook, coauthored by Robert Miller, an important player in the pupfish story as we will see shortly, put it, "Gaps in knowledge, however, make it difficult to achieve a purely ideal aim. The result is that modern classifications of fishes, although based on ideal aims, are always partly artificial." Karl F. Lagler, John E. Bardach, and Robert R. Miller, *Ichthyology* (New York: Wiley, 1962), 9.

50. Carl L. Hubbs, *Studies of the Fishes of the Order Cyprinodontes VI*, University of Michigan Museum of Zoology Miscellaneous Publications 16 (Ann Arbor: University of Michigan, July 9, 1926), 17.

51. For an overview of Hubbs's career, see Elizabeth N. Shor, Richard H. Rosenblatt, and John D. Isaacs, *Carl Leavitt Hubbs, 1894–1979, A Biographical Memoir* (Washington, D.C.: National Academy of Sciences, 1987).

52. Eugenie Clark, *Lady with a Spear* (New York: Harper, 1951), 25–26.

53. Ibid., 26–27.

54. Kenneth S. Norris, "To Carl Leavitt Hubbs, a Modern Pioneer Naturalist on the Occasion of his Eightieth Year," *Copeia* 1974, no. 3 (October 18, 1974):583.

55. Lowell Sumner to Carl Hubbs, August 14, 1951; and Carl Hubbs to Lowell Sumner, August 17, 1951, folder 22, box 81, Hubbs papers, UCSD.

56. Carl Hubbs to Robert R. Miller, March 25, 1937, folder 11, box 230, Hubbs papers, UCSD.

57. Joseph Wales to the Desert Fishes Council, October 21, 1985, folder "*Cyprinodon diabolis*," Taxonomic Files, Miller papers, UMMZ.

58. Joseph H. Wales, "Biometrical Studies of Some Races of Cyprinodont Fishes, from the Death Valley Region, with Description of *Cyprinodon diabolis*, n.sp.," *Copeia* 1930, no. 3 (September 30, 1930):67.

59. Ibid.

60. Ibid., 67–68.

61. Ibid., 68.

62. Carl Hubbs to Joseph Wales, June 17, 1930, folder "*Cyprinodon diabolis*," Taxonomic Files, Miller papers, UMMZ.

63. Joseph H. Wales, "Life History of the Blue Rockfish, *Sebastodes mystinus*" (master's thesis, Stanford University, 1932); Joseph H. Wales obituary, *Corvalis Gazette-Times* (Oregon), August 23, 2002, http://www.gazettetimes.com/joseph-h-wales/article_18a4266f-f2e0-5339-b9b4-6b888396ae60.html.

64. Carl L. Hubbs to Richard S. Corker, March 8, 1937, folder 11, box 230, Hubbs papers, UCSD.

65. Robert Charles Cashner, Gerald R. Smith, Gifford Hubbs Miller, Frances Miller Cashner, and Barry Chernoff, "Robert Rush Miller (1916–2003)," *Copeia* 2011, no. 2 (June 28, 2011):342–347.

66. Robert R. Miller to Carl Hubbs, March 16, 1937, folder 11, box 230, Hubbs papers, UCSD.

67. Robert Rush Miller, Clark Hubbs, and Frances H. Miller, "Ichthyological Exploration of the American West: The Hubbs-Miller Era, 1915–1950," in *Battle Against Extinction: Native Fish Management in the American West*, eds. W. L. Minckley and James E. Deacon (Tucson: University of Arizona Press, 1991), 25.

68. Robert R. Miller, July 5, 1938, vol. "RR Miller 1938," box 25A, field notes collection, University of Michigan Museum of Zoology (hereafter field notes, UMMZ). On Hubbs and Miller, see also Robert J. Edwards, "Carl Leavitt Hubbs and Robert Rush Miller," in *Standing between Life and Extinction: Ethics and Ecology of Conserving Aquatic Species in North American Deserts*, eds. David L. Propst, Jack E. Williams, Kevin R. Bestgen, and Christopher W. Hoagstrom (Chicago: University of Chicago Press, 2020), 13–16.

69. Robert R. Miller, July 14, 1938, vol. "RR Miller 1938," box 25A, field notes, UMMZ.

70. Robert R. Miller, August 15, 1938, vol. "RR Miller 1938," box 25A, field notes, UMMZ.

71. Robert R. Miller, "The Status of *Cyprinodon macularius* and *Cyprinodon nevadensis*, Two Desert Fishes of Western North America," University of Michigan Museum of Zoology Occasional Papers 473 (Ann Arbor: University of Michigan Press, 1943), 1.

72. Robert R. Miller, *The Cyprinodont Fishes of the Death Valley System of Eastern California and Southwestern Nevada*, University of Michigan Museum of Zoology Miscellaneous Publications 68, (Ann Arbor: University of Michigan Press, 1948), 14.

73. Carl L. Hubbs, "Criteria for Subspecies, Species and Genera, as Determined by Researches on Fishes," *Annals of the New York Academy of Sciences* 44, no. 2 (1943):112.

74. Miller, *Cyprinodont Fishes*, 83.

75. Robert Rush Miller, "Records of Some Native Freshwater Fishes Transplanted into Various Waters of California, Baja California, and Nevada," *California Fish and Game* 54, no. 3 (1968):170–179.

76. Miller, *Cyprinodont Fishes*, 111–112.

77. Robert R. Miller to John Kopec, November 2, 1949, folder "Cyprinodon diabolis," Taxonomic Files, Miller papers, UMMZ.

78. Sean C. Lema and Gabrielle A. Nevitt, "Testing an Ecophysiological Mechanism of Morphological Plasticity in Pupfish and Its Relevance to Conservation Efforts for Endangered Devils Hole Pupfish," *Journal of Experimental Biology* 209 (2006):3499–3509; and Sean C. Lema, "Hormones, Developmental Plasticity, and Adaptive Evolution: Endocrine Flexibility as a Catalyst for 'Plasticity-First' Phenotypic Divergence," *Molecular and Cellular Endocrinology* 502 (2020):110678.

79. Miller, *Cyprinodont Fishes*, 85–86. Miller and Hubbs published in several places on the Pleistocene lakes and streams of the Death Valley system, integrating geological and zoological evidence. See also Robert R. Miller, "Correlation between Fish Distribution and Pleistocene Hydrography in Eastern California and Southwestern Nevada, with a Map of the Pleistocene Waters," *Journal of Geology* 54, no. 1 (1946):43–53; Carl L. Hubbs and Robert R. Miller, "The Zoological Evidence: Correlation between Fish Distribution and Hydrographic History in the Desert Basins of the Western United States," in *The Great Basin, with Emphasis on Glacial and Postglacial Times*, Bulletin of the University of Utah 38, no. 20 (1948):17–166; and Robert Rush Miller, "Speciation in Fishes of the Genera *Cyprinodon* and *Empetrichthys*, Inhabiting the Death Valley Region," *Evolution* 4, no. 2 (1950):155–163.

80. Wales to the Desert Fishes Council, October 21, 1985, folder "*Cyprinodon diabolis*," Taxonomic Files, Miller papers, UMMZ.

81. Harry S. Truman, Presidential Proclamation No. 2961, 17 Fed. Reg. 691 (January 23, 1952).

Chapter 2. To Protect and Conserve

1. A much-abbreviated version of this chapter appeared as Kevin C. Brown, "The 'National Playground Service' and the Devils Hole Pupfish," *Forest History Today* (Spring 2017):35–40. On the role of science in shaping the limits for Devils Hole pupfish conservation, see also, Kevin C. Brown "'An Exceedingly Simple, Little Ecosystem': Devils Hole, Endangered Species Conservation, and Scientific Environments," *Notes and Records: The Royal Society Journal of the History of Science* 75, no. 2 (2021):239–257.

2. Gilbert, "Report on the Fishes of the Death Valley Expedition Collected in Southern California and Nevada in 1891, with Descriptions of New Species," 233–234.

3. Miller, *The Cyprinodont Fishes of the Death Valley System of Eastern California and Southwestern Nevada*, 101.

4. Robert Rush Miller, "Status of Populations of Native Fishes of the Death Valley System in California and Nevada," typescript report for National Park Service, August 4, 1967, library, NPS-Cow Creek, 7.

5. While almost nothing is known about the life history or behavior of the species, Miller noted that "*E. merriami* prefers the deeper springs, where it dwells near the bottom"; Miller, *Cyprinodont Fishes*, 101. The decline and extinction of the poolfish may be the result of the introduction of (bottom-dwelling) crayfish to Ash Meadows springs in the 1930s, a possibility Miller later considered. See Robert R. Miller, James D. Williams, and Jack E. Williams, "Extinctions of North American Fishes during the Past Century," *Fisheries* 14, no. 6 (1989):32. The closest living relative of the Ash Meadows poolfish, the Pahrump poolfish (*Empetrichthys latos latos*) has had a population decimated after the introduction of crayfish to one of its habitats, lending credence to this theory. See U.S. Fish and Wildlife Service, "Withdrawal of Proposed Rule to Reclassify the Pahrump poolfish (*Empetrichthys latos*) from Endangered to Threatened Status," 69 Fed. Reg. 17384 (April 2, 2004).

6. Conrad Wirth to Carl P. Russell, November 1, 1943, box 2147, central classified files (CCF), RG 79, NARA-College Park.

7. Mark V. Barrow Jr., *Nature's Ghosts: Confronting Extinction from the Age of Jefferson to the Age of Ecology* (Chicago: University of Chicago Press, 2009), 13.

8. Andrew C. Isenberg, *The Destruction of the Bison: An Environmental History, 1750–1920* (Cambridge: Cambridge University Press, 2000); and Jennifer Price, *Flight Maps: Adventures with Nature in Modern America* (New York: Basic Books, 1999), 1–55.

9. William T. Hornaday, *Our Vanishing Wildlife: Its Extermination and Preservation* (1913; repr. New York: Arno and New York Times, 1970), ix. On changing attitudes toward wildlife in the early twentieth century, see also Thomas R. Dunlap, *Saving America's Wildlife: Ecology and the American Mind, 1850–1990* (Princeton, NJ: Princeton University Press, 1988).

10. Francis B. Sumner, "The Need for a More Serious Effort to Rescue a Few Fragments of Vanishing Nature," *The Scientific Monthly* 10, no. 3 (March 1920):239.

11. Charles T. Brues, "Studies of the Fauna of Hot Springs in the Western United States and the Biology of Thermophilous Animals," *Proceedings of the American Academy of Arts and Sciences* 63, no. 4 (1928):140.

12. Noel M. Burkhead, "Extinction Rates in North American Freshwater Fishes, 1900–2010," *BioScience* 62, no. 9 (2012):798–808. See also Miller, Williams, and Williams, "Extinctions of North American Fishes During the Past Century"; and Jack E. Williams and Donald W. Sada, "Ghosts of Our Making: Extinct Aquatic Species of the North American Desert Region," in Propst, et al., *Standing between Life and Extinction*, 89–105.

13. Robert R. Miller to Carl Hubbs, October 29, 1938, folder 12, box 230, Hubbs papers, UCSD. Miller referred specifically to the Leon Springs pupfish. The species was later rediscovered in 1965 in another spring some 10 kilometers north of Leon Springs. See Stephen E. Kennedy, "Life History of the Leon Springs Pupfish, *Cyprinodon bovinus*," *Copeia*, no. 1 (March 16, 1977):93–103.

14. Robert R. Miller to Carl L. Hubbs, June 22, 1937, folder 11, box 230, Hubbs papers UCSD.

15. George S. Myers to Robert R. Miller, June 30, 1937, regional index boxes, Death Valley system folder, Miller papers, UMMZ.

16. Joseph Wales to Robert R. Miller, taxonomic files, *C. diabolis* folder, April 27, 1938, Miller papers, UMMZ.

17. Robert R. Miller to Carl Hubbs, March 16, 1937, and Carl Hubbs to Robert R. Miller, March 25, 1937, folder 11, box 230, Hubbs papers, UCSD.

18. Peter S. Alagona, *After the Grizzly: Endangered Species and the Politics of Place in California* (Berkeley: University of California Press, 2013), 50. On Grinnell and "salvage collecting," see also Kohler, *All Creatures*, 147–149.

19. James Delbourgo, *Collecting the World: Hans Sloane and the Origins of the British Museum* (Cambridge, Mass.: The Belknap Press of Harvard University Press, 2017); and Asma, *Stuffed Animals and Pickled Heads.*

20. Hilda Wood Grinnell, "Joseph Grinnell: 1877–1939," *The Condor* 42, no. 1 (1940):3–34; and Alagona, *After the Grizzly*, 45–52. Joseph Grinnell was not related to naturalist George Bird Grinnell.

21. Joseph Grinnell, "The Museum Conscience," *Museum Work* 4 (September–October 1921):63.

22. Scientists like Grinnell also played a role in the establishment of state fish and game laws during the early twentieth century that worked to halt the decline in animals sought for meat, fur, or feathers. See Alagona, *After the Grizzly*, 57–68.

23. On the Ecological Society of America, see Sara Fairbank Tjossem, "Preservation of Nature and Academic Respectability: Tensions in the Ecological Society of America, 1915–1979" (PhD diss., Cornell University, 1994); and Alagona, *After the Grizzly*, 71–76.

24. Joseph Grinnell and Tracy I. Storer, "Animal Life as an Asset of National Parks," *Science* 44, no. 1133 (September 15, 1916):379.

25. T.S. Palmer, "National Monuments as Wild-Life Sanctuaries," in *Proceedings of the National Parks Conference* (Washington, D.C.: GPO, 1917), 225; and Hal Rothman, *Preserving Different Pasts: The American National Monuments* (Urbana: University of Illinois Press, 1989), 94–97. A word on the difference between national parks and national monuments: NPS manages national parks and many, but not all, national monuments. Generally, parks are designated by Congress while monuments originate with executive action by the president. Especially since the 1930s, NPS has managed both under the same broad principles. Further blurring this distinction, many monuments—including Death Valley and Grand Canyon—have later been redesignated by Congress as national parks. For the purposes of the pupfish, the significant issues did not revolve around whether Devils Hole become part of a national park or a national monument but whether it would be managed by the National Park Service at all. On the history of national monuments, see Rothman, *Preserving Different Pasts.*

26. Sumner, "The Need for a More Serious Effort," 241.

27. Ibid., 246. See also Francis B. Sumner, "The Responsibility of the Biologist in the Matter of Preserving Natural Conditions," *Science* 54, no. 1385 (July 15, 1921):39–43.

28. Miller provided recommendations for locations and collecting methods. Robert R. Miller to F.B. Miller, March 25, 1939, Acc. 1939-X:2, accession records, University of Michigan Museum of Zoology. F.B. Sumner and M.C. Sargent, "Some Observations on the Physiology of Warm Spring Fishes," *Ecology* 21, no. 1 (January 1940):45–54. Sumner and Sargent visited Devils Hole but did not use *C. diabolis* in the experiment. On the history of the discipline of ecology, see Sharon E. Kingsland, *The Evolution of American Ecology, 1890–2000* (Baltimore: Johns Hopkins University Press, 2005).

29. F.B. Sumner to Carl Hubbs, November 19, 1941, folder 15, box 230, Hubbs papers, UCSD.

30. When F.B. Sumner pitched his idea in 1941, Hubbs told him he had "given considerable thought" to protecting isolated fishes from development but admitted having done little

to advocate for habitat protection. Hubbs promised Sumner that he and Miller would "pay particular attention to this question of preservation" during their next trip to the West and be ready to list the "most desirable" for protection by late 1942 or early 1943. Carl Hubbs to F. B. Sumner, November 25, 1941, folder 15, box 230, Hubbs papers, UCSD. Hubbs had recommended protecting some habitat for fishes in Texas in a 1940 article. See Carl L. Hubbs, "Fishes from the Big Bend Region of Texas," *Transactions of the Texas Academy of Science* 23 (1940):3–12.

31. Carl Hubbs to Carl P. Russell, October 21, 1943, box 2147, central classified files (CCF), RG 79, NARA-College Park.

32. Carl L. Hubbs and Orthello L. Wallis, "The Native Fish Fauna of Yosemite National Park and its Preservation," *Yosemite Nature Notes* 27, no. 12 (December 1948):131–144; and Carl L. Hubbs and Karl F. Lagler, "Fishes of Isle Royale, Lake Superior, Michigan," *Papers of the Academy of Science, Arts, and Letters* 33 (1947; University of Michigan Press, 1949):73–133.

33. On fish stocking in Yosemite, see Carl Hubbs to Victor Cahalane, September 3, 1943, folder 21, box 81, Hubbs papers, UCSD.

34. Carl Hubbs to Newton B. Drury, November 22, 1943, folder 21, box 81, Hubbs papers, UCSD.

35. Newton B. Drury to Carl Hubbs, January 9, 1950, folder 22, box 81, Hubbs papers, UCSD.

36. Clifford Presnall to Carl Russell, no date, circa November 1943, box 2147, central classified files (CCF), RG 79, NARA-College Park.

37. Conrad Wirth to Carl P. Russell, November 1, 1943, box 2147, CCF, RG 79, NARA-College Park.

38. Victor Cahalane to Carl P. Russell, December 21, 1943, box 2147, CCF, RG 79, NARA-College Park.

39. Carl P. Russell to Carl Hubbs, December 21, 1943, box 2147, CCF, RG 79, NARA-College Park. Historian Lary Dilsaver has uncovered documents in the NPS Harpers Ferry, West Virginia, archives showing that in 1943 the agency added "Ash Meadow Desert Springs" in Inyo County, California, to a list of areas proposed for NPS protection. Despite the inaccuracy of the Ash Meadows name ("Meadows" is plural) and location (actually in Nye County, Nevada), the addition was in response to Hubbs's 1943 letter. Lary M. Dilsaver, "Not of National Significance: Failed National Park Proposals in California," *California History* 85, no. 2 (2008):7; and Lary Dilsaver, email message to author, November 14, 2019.

40. National Park Service Organic Act, Pub. L. 64-235, 39 Stat. 535 (1916). On early Park Service history, see Alfred Runte, *National Parks: The American Experience*, 3rd ed. (Lincoln: University of Nebraska Press, 1997); Barry Mackintosh, *The National Parks: Shaping the System* (Washington, D.C.: U.S. Department of the Interior, 1985); Robert B. Keiter, *To Conserve Unimpaired: The Evolution of the National Park Idea* (Washington, D.C.: Island Press, 2013); and Richard West Sellars, *Preserving Nature in the National Parks: A History* (New Haven, CT: Yale University Press, 1997).

41. Franklin K. Lane to Stephen T. Mather, May 13, 1918, in *America's National Park System: The Critical Documents*, ed. Lary M. Dilsaver, 2nd ed. (New York: Rowman and Littlefield, 2016), 36, 38–39. A powerful literature has developed showing how the NPS vision of wild "playgrounds" variously excluded or ignored Native Americans, whose ancestors had been living in the parks for generations. Biologists' focus on the nature of the parks sometimes reinforced the invisibility of Native Americans in supposedly people-less places. See Philip Burnham, *Indian Country, God's Country: Native Americans and the National Parks* (Washington, D.C.: Island Press, 2000); Karl Jacoby, *Crimes Against Nature: Squatters, Poachers, Thieves, and the Hidden*

History of American Conservation (Berkeley: University of California Press, 2001); and Mark David Spence, *Dispossessing the Wilderness: Indian Removal and the Making of the National Parks* (New York: Oxford University Press, 1999).

42. Proclamation No. 2028, 47 Stat. 2554 (1933); and Rothman and Miller, *Death Valley National Park*, 28–33.

43. National Park Service, *Death Valley National Monument, California* (Washington, D.C.: GPO, 1934), 1, https://www.nps.gov/parkhistory/online_books/brochures/1934/deva/1934.pdf.

44. Sumner, "The Need for a More Serious Effort," 246.

45. George M. Wright, Joseph S. Dixon, and Ben H. Thompson, *Fauna of the National Parks of the United States: A Preliminary Survey of Faunal Relations in National Parks*, Fauna Series 1 (Washington, D.C.: GPO, 1933), 147–148.

46. Sellars, *Preserving Nature in the National Parks*, 99, 101, 167, 91–148. See also, Lowell Sumner, "Biological Research in the National Park Service: A History," *George Wright Forum* 3, no. 4 (1983):3–27; and James A. Pritchard, "The Meaning of Nature: Wilderness, Wildlife, and Ecological Values in the National Parks," *George Wright Forum* 19, no. 2 (2002):46–56.

47. Sellars, *Preserving Nature in the National Parks*, 164.

48. Lowell Sumner to Carl Hubbs, August 14, 1951, folder 22, box 81, Hubbs papers, UCSD.

49. Carl Hubbs to J. Clark Salyer II, September 12, 1949, folder 22, box 81, Hubbs papers, UCSD.

50. U.S. Department of the Interior, *Annual Report of the Secretary of the Interior, Fiscal Year Ending June 30, 1946* (Washington, D.C.: GPO, 1947), 278. On the history of the Fish and Wildlife Service, see Nathaniel P. Reed and Dennis Drabelle, *The United States Fish and Wildlife Service* (Boulder, CO: Westview Press, 1984); and Charles G. Curtin, "The Evolution of the U.S. National Wildlife Refuge System and the Doctrine of Compatibility," *Conservation Biology* 7, no. 1 (March 1993):29–38.

51. U.S. Department of the Interior, *Annual Report… 1946*, 277–278.

52. "Editorial News and Notes," *Copeia* 1946, no. 3 (October 20, 1946):180.

53. Robert P. Davison, Alessandra Falucci, Luigi Maiorano, and J. Michael Scott, "The National Wildlife Refuge System," in *The Endangered Species Act at Thirty*, vol. 1, *Renewing the Conservation Promise*, eds. Dale D. Goble, J. Michael Scott, and Frank W. Davis (Washington, D.C.: Island Press, 2005), 90–100.

54. Robert M. Wilson, *Seeking Refuge: Birds and Landscapes of the Pacific Flyway* (Seattle: University of Washington Press, 2010), 34.

55. Ibid., 99.

56. Doren Woodward, "Preliminary Reconnaissance of Ash Meadows, Nevada," 1947, box 62, P236, RG 22, NARA-College Park.

57. Frank W. Groves to Regional Director, Portland, FWS, April 4, 1947, box 62, P236, RG 22, NARA-College Park.

58. Wilson, *Seeking Refuge*, 131.

59. A.J. Rissman to Regional Director, Portland, May 21, 1947, box 62, P236, RG 22, NARA-College Park.

60. Hubbs and Salyer even cowrote a report on the management of Iowa's fisheries before Salyer joined the Biological Survey. See Carl L. Hubbs and J. Clark Salyer, "Fish Conditions in Iowa Lakes, with Recommendations for the Fish Management of these Lakes," Report 209 (Ann Arbor: Institute for Fisheries Research, Michigan Department of Conservation and University of Michigan, May 2, 1933). On Salyer, see also "J. C. Salyer, Conservationist," obituary, *Washington Post*, August 18, 1966, B6.

61. J. Clark Salyer II to William M. Rush, April 14, 1937, box 246, entry 162, RG 22, NARA-College Park.

62. J. Clark Salyer II, handwritten note, bottom of May 16, 1947, A. J. Rissman to J. Clark Salyer, box 62, P236, RG 22, NARA-College Park.

63. J. Clark Salyer to Carl Hubbs, December 5, 1950, folder 7, box 238, Hubbs papers, UCSD.

64. Research on the plants of Ash Meadows developed later, especially during the 1970s and 1980s. See Sada, *Recovery Plan for the Endangered and Threatened Species of Ash Meadows, Nevada*, 22–27.

65. Robert R. Miller to Carl Hubbs, June 22, 1937, folder 11, box 230, Hubbs papers, UCSD; and Robert R. Miller to George S. Myers, June 24, 1937, regional index boxes, Death Valley system folder, Miller papers, UMMZ. These specimens are catalog number 132913, division of fishes, University of Michigan Museum of Zoology (accessed through Fishnet2 Portal, www.fishnet2.net).

66. Wales, "Biometrical Studies," 61–62.

67. Myers, accession memorandum, accession 131034, September 18, 1934, fishes division, National Museum of Natural History, Smithsonian Institution. Specimens are catalog number 94266.50309, fishes division, National Museum of Natural History, Smithsonian Institution (accessed through Fishnet2 Portal, www.fishnet2.net).

68. James E. Deacon, interview by Kevin C. Brown, Henderson, Nevada, April 3, 2014, transcript, author's possession, 26.

69. Carl Hubbs to Robert R. Miller, April 8, 1937, folder 11, box 230, Hubbs papers, UCSD.

70. Tamar Dayan and Bella Galil, "Natural History Collections as Dynamic Research Archives," in *Stepping in the Same River Twice: Replication in Biological Research*, eds. Ayelet Shavit and Aaron M. Ellison (New Haven, CT: Yale University Press, 2017), 55–79.

71. Sanford R. Wilbur, *The California Condor, 1966–76: A Look at Its Past and Future*, North American Fauna 72 (Washington, D.C.: U.S. Department of the Interior, 1978), 20–21. See also Barrow, *Nature's Ghosts*, 285–286; and Alagona, *After the Grizzly*, 127.

72. On the great auk, see Barrow, *Nature's Ghosts*, 61–66; and Jessica E. Thomas et al., "Demographic Reconstruction from Ancient DNA Supports Rapid Extinction of the Great Auk," *eLife* 8 (2019):e47509, doi:10.7554/eLife.47509.

73. Ben A. Minteer, James P. Collins, Karen E. Love, and Robert Puschendorf, "Avoiding (Re)extinction," *Science* 344 (April 18, 2014):260–261, doi:10.1126/science.1250953; and Ricardo Rodríguez-Estrella and Ma Carmen Blázquez Moreno, "Rare, Fragile Species, Small Populations, and the Dilemma of Collections," *Biodiversity and Conservation* 15 (2006):1621–1625, doi:10.1007/s10531-004-4308-6.

74. Results of "taxon" search for "Cyprinodon diabolis," Fishnet2 Portal, www.fishnet2.net. Miller incorporated almost all of these specimens into his study, utilizing 195 *C. diabolis* specimens in his studies. See Miller, *Cyprinodont Fishes*, 83.

75. Devils Hole pupfish adult population count data, 1972–2015, Death Valley National Park, National Park Service, in author's possession. Before groundwater pumping began in the 1960s, the water level in Devils Hole was higher than it has been since. With more habitat area, it is possible that pupfish populations were regularly higher than scuba estimates have recorded.

76. Robert R. Miller, field notes, April 1, 1950, box 29A, field notes, UMMZ.

77. Ira La Rivers, *Fishes and Fisheries of Nevada* (Nevada State Fish and Game Commission, 1962; repr., Reno: University of Nevada Press, 1994). On La Rivers, see Gary Vinyard and James E. Deacon, foreword to 1994 edition of *Fishes and Fisheries of Nevada* (Reno: University of Nevada

Press, 1994), 1–4; Thomas P. Lugaski, "Obituary: Ira John La Rivers, II, 1915–1977," *Pan-Pacific Entomologist* 55, no. 3 (1979):230–233; and Hany Abdoun, "Ira John La Rivers (1915–1977)," California Academy of Sciences, http://researcharchive.calacademy.org/research/library/special/bios/LaRivers.pdf.

78. Robert R. Miller to Ira La Rivers, October 13, 1952, folder 4, box 239, Hubbs papers, UCSD.

79. Devils Hole pupfish specimens collected April 10, 1950, by La Rivers and assistants are specimen collections 19160 and 19161 in the Museum of Natural History, University of Nevada, Reno. For catalogue of these and all other University of Nevada, Reno, Devils Hole pupfish specimens, see Arctos Collaborative Collection Management Solution, https://arctos.database.museum.

80. Ira La Rivers, "The Dryopoidea Known or Expected to Occur in the Nevada Area (Coleoptera)," *Wasmann Journal of Biology* 8, no. 1 (Spring 1950):106; and Ira La Rivers and T.J. Trelease, "An Annotated Checklist of the Fishes of Nevada," *California Fish and Game* 38, no. 1 (1952):119.

81. La Rivers, "The Dryopoidea," 106; and La Rivers and Trelease, "An Annotated Checklist," 119.

82. Ira La Rivers collection, Library and Archives, California Academy of Sciences, San Francisco, CA.

83. Ira La Rivers to Robert R. Miller, October 16, 1958, Taxonomic Files, folder *Cyprinodon diabolis*, Miller papers, UMMZ.

84. Christopher H. Martin, Jacob E. Crawford, Bruce J. Turner, and Lee H. Simons. "Diabolical Survival in Death Valley: Recent Pupfish Colonization, Gene Flow, and Genetic Assimilation in the Smallest Species Range on Earth," *Proceedings of the Royal Society B: Biological Sciences* 283, no. 1823 (2016):4–5, doi:10.1098/rspb.2015.2334.

85. La Rivers and Trelease, "An Annotated Checklist," 119.

86. Carl Hubbs to Newton Drury, November 27, 1950, box 1622, P11, RG 79, NARA-College Park.

87. Carl Hubbs to Ira La Rivers, January 15, 1952, folder 7, box 238, Hubbs papers, UCSD.

88. Carl Hubbs to E. E. Ogston, July 7, 1953, folder 7, box 238, Hubbs papers, UCSD.

89. On the history of the Bureau of Land Management, see James R. Skillen, *The Nation's Largest Landlord: The Bureau of Land Management in the American West* (Lawrence: University Press of Kansas, 2009).

90. Jean M.F. Dubois, "Report: Devils Hole Proposed Withdrawal," 2–3, SF 94949, box 2108, P11, RG 79, NARA-College Park.

91. Newton B. Drury to Region 4 Director, NPS, June 23, 1950, box 2108, P11, RG 79, NARA-College Park.

92. T.R. Goodwin to Regional Director, NPS, July 25, 1950, box 2108, P11, RG 79, NARA-College Park.

93. Floyd Keller to T.R. Goodwin, September 5, 1950, box 2108, P11, RG 79, NARA-College Park.

94. Newton B. Drury to Director, BLM, October 12, 1950, box 2108, P11, RG 79, NARA-College Park.

95. See Carl Hubbs to David Soltz, March 16, 1979, folder 10, box 50, Hubbs papers, UCSD; and Carl Hubbs, digest of letters in Death Valley Monument files on proposed areas, circa November 1950, folder 1, box 238, Hubbs papers, UCSD.

96. Carl Hubbs to Newton B. Drury, November 27, 1950, box 1622, P11, RG 79, NARA-College Park.

97. Carl Hubbs to Floyd Keller, December 13, 1950, folder 22, box 81, Hubbs papers, UCSD.

98. J. Clark Salyer to Carl Hubbs, December 5, 1950, folder 7, box 238, Hubbs papers, UCSD.

99. J. Clark Salyer to Ira Gabrielson, December 5, 1950; J. Clark Salyer to Howard Zahniser, December 5, 1950; and J. Clark Salyer to Carl Hubbs, telegram, January 18, 1951, all in folder 7, box 238, Hubbs papers, UCSD.

100. Lawrence C. Merriam to Director, NPS, January 17, 1951, box 1622, P11, RG79, NARA-College Park.

101. On Sumner's career, see Lowell Sumner, "Biological Research in the National Park Service."

102. Lowell Sumner, "A Pictorial Summary of Devils Hole, Nevada and Recommendations," January 1951, box 1622, P11, RG79, NARA-College Park.

103. Carl Hubbs to Newton B. Drury, November 27, 1950, box 1622, P11, RG 79, NARA-College Park.

104. Sumner, "A Pictorial Summary of Devils Hole, Nevada and Recommendations."

105. In the winter of 1950–1951, the Park Service was in a state of flux, as the agency's director, Newton Drury, resigned over a plan to build the Echo Park Dam in the Colorado portion of Dinosaur National Monument, even though his boss, the Secretary of the Interior, supported it. See Sellars, *Preserving Nature in the National Parks*, 178; and Mark W. T. Harvey, *A Symbol of Wilderness: Echo Park and the American Conservation Movement* (Albuquerque: University of New Mexico Press, 1994), 101–102.

106. Antiquities Act of 1906, Pub. L. 59-209, 34 Stat. 225 (1906). On the Antiquities Act, see Rothman, *Preserving Different Pasts*; David Harmon, "The Antiquities Act and Nature Conservation," in *The Antiquities Act: A Century of American Archaeology, Historical Preservation, and Nature Conservation*, eds. David Harmon, Francis P. McManamon, and Dwight T. Pitcaithley (Tucson: University of Arizona Press, 2006), 199–215; Ronald F. Lee, *The Antiquities Act of 1906* (Washington, D.C.: National Park Service, 1970); and Congressional Research Service, *National Monuments and the Antiquities Act*, CRS Report R41330 (Congressional Research Service, November 30, 2018).

107. Arthur Demaray to Region 4 Director, NPS, March 12, 1951, box 1622, P11, RG 79, NARA-College Park.

108. Proclamation No. 2961, 17 Fed. Reg. 691 (Jan. 23, 1952).

109. Acting Secretary of the Interior, R. D. Searles, submitted a draft proclamation to President Truman for the addition of Devils Hole to Death Valley National Monument in December 1951 (box 1622, P11, RG 79, NARA-College Park). The final version signed by Truman does not differ in substance, though it does have minor textual changes (see Proclamation No. 2961). For Truman's daily calendar from January 17, 1952 (the date he signed the proclamation), see: https://www.trumanlibrary.gov/calendar?month=1&day=17&year=8.

110. Lowell Sumner to Carl Hubbs, August 14, 1951, folder 22, box 81, Hubbs papers, UCSD.

111. Dubois, "Report: Devils Hole Proposed Withdrawal," 1.

112. Carl Hubbs to Ronald Lee, January 29, 1952, L58, central files, NPS-Cow Creek.

113. On forgetting Devils Hole in the annual report, see E. E. Ogston to Conrad Wirth, May 30, 1953, box 108, P11, RG 79, NARA-College Park. On request to keep Devils Hole off of maps, see Lawrence C. Merriam to Automobile Club of Southern California, and Lawrence C. Merriam to Rand McNally and Company, both March 10, 1952, box 1622, P11, RG 79, NARA-College Park.

114. Death Valley did install a gate at the easiest passage down to Devils Hole in 1953 after Carl Hubbs complained to NPS's Washington staff, but this gate was easily circumvented. For full NPS accounting of the incident and the rescue and recovery effort related to the deaths of

David Rose and Paul Giancontieri in Devils Hole on the night of June 20–21, 1965, see box 607, P11, RG 79, NARA-College Park.

115. Edwin C. Kenner to Regional Director, Western Region, February 8, 1965, box 2301, P11 RG 79, NARA-College Park.

116. Chief Naturalist, monthly report, March 1953, K2615, NPS-Cow Creek.

Chapter 3. Beneficial Use

1. Robert R. Miller, field notebook, July 26, 1938, vol. "RR Miller 1938," box 25A, field notes, UMMZ.

2. An even smaller circle may be those who have had nightmares about blowing deadlines for book manuscripts about the history of pupfish.

3. Carl Hubbs to Robert R. Miller, April 8, 1937, folder 11, box 230, Hubbs papers, UCSD.

4. John Wesley Powell, *Report on the Lands of the Arid Region of the United States*, 2nd ed. (Washington, D.C.: GPO, 1879), esp. 6–9. On Powell, see Stegner, *Beyond the Hundredth Meridian*.

5. Walter Prescott Webb, "The American West, Perpetual Mirage," *Harper's Magazine*, May 1, 1957, 26.

6. On water and the West see, Donald Worster, *Rivers of Empire: Water, Aridity, and the Growth of the American West* (New York: Oxford University Press, 1985); Donald J. Pisani, *Water, Land, and Law in the West: The Limits of Public Policy, 1850–1920* (Lawrence: University Press of Kansas, 1996); Norris Hundley Jr., *The Great Thirst: Californians and Water: A History*, rev. ed. (Berkeley: University of California Press, 2001); Marc Reisner, *Cadillac Desert: The American West and Its Disappearing Water*, rev. ed. (New York: Penguin Books, 1993); and B. Lynn Ingram and Frances Malamud-Roam, *The West without Water: What Past Floods, Droughts, and Other Climatic Clues Tell Us about Tomorrow* (Berkeley: University of California Press, 2013). On species and water in the West, see Alagona, *After the Grizzly*; and Joseph E. Taylor III, *Making Salmon: An Environmental History of the Northwest Fisheries Crisis* (Seattle: University of Washington Press, 1999).

7. Hundley, *The Great Thirst.*

8. W. Michael Hanemann, "The Central Arizona Project," Department of Agricultural and Resource Economics and Policy, Division of Agricultural and Natural Resources, University of California at Berkeley, Working Paper 937, October 2002, https://escholarship.org/uc/item/87k436cf.

9. NOAA National Centers for Environmental Information, State Summaries, Nevada, 149-NV, https://statesummaries.ncics.org/nv.

10. Michael S. Green, *Nevada: A History of the Silver State* (Reno: University of Nevada Press, 2015), 198–202; and Donald J. Pisani, "Federal Reclamation and Water Rights in Nevada," *Agricultural History* 51, no. 3 (1977):540–558.

11. KNBC Television, "Timetable for Disaster," 1970, transcript, 13, FWS-Las Vegas.

12. John Bradford, "Final Proof," July 31, 1916, Land Entry Case File, Patent 558456, RG 49, NARA-Online. Searches of the Bureau of Land Management General Land Office database (https://glorecords.blm.gov) reveal more than a dozen land patents for Homestead Act entries granted between 1910 and 1930 on the four townships centered around Ash Meadows (T17-18S, R50-51E, Mount Diablo Meridian).

13. Letters of Coleman and Diehl in Application 2890, Nevada Division of Water Resources, office records, Carson City, Nevada (hereafter DWR-Carson City). Newspaper references to James Coleman as a prospector found in the *Tonopah Bonanza* from 1900s and 1910s, via

searches of the Library of Congress's Chronicling America newspaper archive. Charles Diehl was an assayer in Goldfield during 1907–1910, according to "Fails to Locate Bar in Tonopah," *Tonopah Times-Bonanza*, July 15, 1955, 12; cited in Michael Hodder, "Western American Gold and Unparted Bars: A Review of the Evidence," *American Journal of Numismatics* 11 (1999): 91–92.

14. John Bradford, "Final Proof," July 31, 1916.

15. Water permit application 2890, DWR-Carson City; and U.S. Bureau of Labor Statistics, "Consumer Price Index Inflation Calculator," https://www.bls.gov/data/inflation_calculator.htm. Five thousand dollars in February 1914 is $130,645.45 in February 2020.

16. John Bradford, "Affidavit of Labor and Improvements," September 24, 1914, Application 2890, DWR-Carson City.

17. By the mid-1950s, it also occurred to several others that Devils Hole could make a promising irrigation source, including Curtis Springer, a patent medicine huckster, radio evangelist, and founder of the Zzyzx health spa ("the last word in health") in the Mojave Desert. Neither Springer nor anyone else got even as far as Bradford's forgotten ditch. See permit application 17009, August 7, 1956, DWR-Carson City. There is no evidence that Hubbs, Miller, La Rivers, or the National Park Service were aware of any of these early attempts to appropriate water from Devils Hole.

18. Joseph Wales to Desert Fishes Council, October 21, 1985; and John Bradford to Joseph Wales, May 26, 1930, both in folder "*Cyprinodon diabolis*," Taxonomic Files, Miller papers. Wales also examined pupfish living in "Bradford Spring" on the Bradford ranch in his 1930 paper to show how the Devils Hole population was unique by comparison.

19. The literature addressing the development of the administrative state apparatus in the United States during the late nineteenth and early twentieth century is a large one that crosses several fields. See especially Samuel P. Hays, *Conservation and the Gospel of Efficiency: The Progressive Conservation Movement, 1890–1920* (Cambridge, MA.: Harvard University Press, 1959); and Stephen Skowronek, *Building a New American State: The Expansion of National Administrative Capacities, 1877–1920* (Cambridge: Cambridge University Press, 1982).

20. Second-feet was the term on the 1914 form. The usage has since been replaced by cubic feet per second (cfs) to express volumetric flow.

21. After passing the 1903 law, the legislature amended it in 1905, and then repealed and replaced it with other laws in 1907 and 1913. On the court cases, laws, and agreements that have shaped water rights in Nevada, see Sylvia Harrison, "The Historical Development of Nevada Water Law," *University of Denver Water Law Review* 5, no. 1 (2001):148–182; and Fred W. Weldon, "History of Water Law in Nevada and the Western States," Background Paper 03-2, Legislative Counsel Bureau, Nevada Legislature, 2003, https://www.leg.state.nv.us/Division/Research/Publications/Bkground/BP03–02.pdf.

22. Water law in the Eastern U.S. has generally followed the English common law–based "riparian" doctrine, which provides landowners adjacent to a body of water the right to its use.

23. Nevada Revised Statutes 533.035, https://www.leg.state.nv.us/NRS/NRS-533.html#NRS533Sec035.

24. For Richard G. Miller resumes and work history, see folders 4 and 15, box 55, Foresta Institute for Ocean and Mountain Studies records, 2001-01, Special Collections, University Libraries, University of Nevada, Reno (hereafter Foresta records, UNR).

25. Richard G. Miller, notes on collecting and rearing Devils Hole pupfish, 1948–1949, folder 22, box 14, Foresta records, UNR.

26. Richard G. Miller to Joseph Wales, September 14, 1955, folder 8, box 14, Foresta records, UNR.

27. Permit application 12781 (Richard G. Miller), December 29, 1948, DWR-Carson City.

28. Hugh A. Schamberger to Ed S. Giles, May 10, 1949, permit application 12781, DWR-Carson City.

29. Hugh A. Schamberger to Ed S. Giles, May 10, 1949, permit application 12781, DWR-Carson City.

30. I have been unable to locate files associated with Jean M. F. DuBois's 1950 land examination report on Devils Hole in BLM archival collections at the National Archives in San Bruno or in agency holdings, despite substantial digging. Therefore, I do not have direct evidence from BLM on what exactly sparked its report to NPS on Devils Hole. I located a copy of the report itself in NPS's records at the National Archives. See Jean M. F. Dubois, "Report: Devils Hole Proposed Withdrawal," SF 94949, box 2108, P11, RG 79, NARA-College Park.

31. Hugh A. Schamberger to Ed S. Giles, May 10, 1949, permit application 12781, DWR-Carson City. In 1953, after Devils Hole had been added to Death Valley National Monument, Schamberger arranged for Miller to withdraw his pending water application. Richard G. Miller to Hugh A. Schamberger, February 28, 1953, permit application 12781, DWR-Carson City.

32. Virginia L. McGuire, *Water-Level and Recoverable Water in Storage Changes, High Plains Aquifer, Predevelopment to 2015 and 2013–15*, Scientific Investigations Report 2017-5040 (Reston, VA: U.S. Geological Survey, 2017).

33. Keith J. Halford and Tracie R. Jackson, *Groundwater Characterization and Effects of Pumping in the Death Valley Regional Groundwater Flow System, Nevada and California, with Special Reference to Devils Hole*, Professional Paper 1863 (Reston, VA: U.S. Geological Survey, 2020), 136–137.

34. Riggs and Deacon, "Connectivity in Desert Aquatic Ecosystems: The Devils Hole Story," 1.

35. Edward Abbey, *The Journey Home: Some Words in Defense of the American West* (New York: Dutton, 1977), 82.

36. Donald E. Green, *Land of the Underground Rain: Irrigation on the High Texas Plains, 1910–1970* (Austin: University of Texas Press, 1973); and Donald Worster, *Dustbowl: The Southern Plains in the 1930s* (New York: Oxford University Press, 1979).

37. Cheryl A. Dieter, Molly A. Maupin, Rodney R. Caldwell, Melissa A. Harris, Tamara I. Ivahnenko, John K. Lovelace, Nancy L. Barber, and Kristin S. Linsey, *Estimated Use of Water in the United States in 2015*, Circular 1441 (Reston, VA: U.S. Geological Survey, 2018), 28, 18. See also Robert Glennon, *Water Follies: Groundwater Pumping and the Fate of America's Fresh Waters* (Washington, D.C.: Island Press, 2002).

38. McGuire, *Water-Level and Recoverable Water in Storage Changes, High Plains Aquifer, Predevelopment to 2015 and 2013–15*, 2; and Laura Parker, "What Happens to the U.S. Midwest When the Water's Gone?" *National Geographic*, August 2016, https://www.nationalgeographic.com/magazine/2016/08/vanishing-midwest-ogallala-aquifer-drought/.

39. Robert R. Miller, "*Rhinichthys deaconi*, a New Species of Dace (Pisces: Cyprinidae) from Southern Nevada," *Occasional Papers of the Museum of Zoology, University of Michigan* 707 (July 24, 1984); and James R. Harrill, *Pumping and Ground-Water Storage Depletion in Las Vegas Valley, Nevada, 1955–74*, Water-Resources Bulletin 44 (State of Nevada Department of Conservation and Natural Resources, Division of Water Resources, 1976), 37–43.

40. U.S. Bureau of Reclamation, *Interim Report on Inland Basins Projects, Nevada-California Amargosa Project, Reconnaissance Investigation*, Region 3 (Boulder City, NV: U.S. Bureau of Rec-

lamation, July 1968), 3. On limited pre-1960s development of groundwater in the Amargosa Desert, see also George E. Walker and Thomas E. Eakin, *Geology and Ground Water of Amargosa Desert, Nevada–California*, Ground-Water Resources—Reconnaissance Series, Report 14 (Carson City: Nevada Department of Conservation and Natural Resources, March 1963), 28; and State of Nevada Division of Water Resources online hydrographic abstract data for Amargosa Desert Basin (230), http://water.nv.gov/hydrographicabstract.aspx.

41. Search conducted through Nevada Division of Water Resources "well log search" (http://water.nv.gov/WellLogQuery.aspx), April 16, 2019.

42. Hugh A. Schamberger, *Evolution of Nevada's Water Laws, as Related to the Development and Evolution of the State's Water Resources, from 1866 to about 1960*, Water-Resources Bulletin 46 (U.S. Geological Survey and Nevada Division of Water Resources, 1991), 33.

43. Daniel Rothberg, "After Nearly Two Decades of Drought, State Engineer Tackles Excess Water Rights and Faces a Backlash in the Courts," *Nevada Independent*, August 6, 2018, https://thenevadaindependent.com/article/after-nearly-two-decades-of-drought-state-engineer-tackles-excess-water-rights-and-faces-a-backlash-in-the-courts.

44. Henry Brean, "Nevada Bans New Residential Wells in Pahrump over Groundwater Decline," *Las Vegas Review-Journal*, December 21, 2017, https://www.reviewjournal.com/news/politics-and-government/nevada/nevada-bans-new-residential-wells-in-pahrump-over-groundwater-decline/.

45. George Swink to Hugh Schamberger, November 29, 1958, folder "Amargosa Drainage Basin—Ash Meadows Correspondence," DWR-Carson City.

46. George Swink to George Hennan, December 11, 1965, permit application 19831, DWR-Carson City.

47. George Swink, World War I Draft Registration Card, June 5, 1917, National Archives and Records Administration microform publication M1509, accessed from familysearch.org database; and Iva M. McFadden, "It Is Our Treat Again: Imperial Heaps American Breakfast Tables with Cantaloupes Only to Face Another Price Debacle," *Los Angeles Times*, July 12, 1931, J3.

48. "Growers Push Fight on Reds," *Los Angeles Times*, March 16, 1934; and Carey McWilliams, *Factories in the Field: The Story of Migratory Farm Labor in California* (1939; repr., Berkeley: University of California Press, 1999), 224–226. See also, Richard A. Walker, *The Conquest of Bread: 150 Years of Agribusiness in California* (New York: New Press, 2004); and Benny J. Andrés Jr., *Power and Control in the Imperial Valley: Nature, Agribusiness, and Workers on the California Borderland, 1900–1940* (College Station: Texas A&M University Press, 2015). Groundwater pumping did not play an important part of irrigation systems in the Imperial Valley during this period. See O.J. Loeltz, Burdge Irelan, J.H. Robison, and F.H. Olmsted, *Geohydrologic Reconnaissance of the Imperial Valley, California*, U.S. Geological Survey Professional Paper 486-K (Washington, D.C.: GPO, 1975):K23–24.

49. City directories from the 1960s and 1970s list Swink's residence on Paseo Del Ocaso, La Jolla, California. Ancestry.com, *U.S. City Directories, 1822–1995*, online database, Provo, UT: Ancestry.com Operations Inc., 2011.

50. George Swink to George Hennan, December 11, 1965, Nye County Land Company permit application 19831, DWR-Carson City.

51. Nevada Test Site Community Act of 1963, S. 2030, 88th Cong. (1963); and Nevada Test Site Community Act of 1963, H.R. 8003, 88th Cong. (1963).

52. *Nevada Test Site Community Act of 1963: Hearings on H.R. 8003 and S. 2030, Part 1, Before the Subcommittee on Legislation and Subcommittee on Communities of the Joint Committee on*

Atomic Energy, 88th Cong. (1963), 125–127. USGS hydrologist Isaac Winograd even produced a report on the hydrological impacts of the project at the different proposed sites, including Ash Meadows. See Isaac J. Winograd, *A Summary of the Ground-Water Hydrology of the Area between the Las Vegas Valley and the Amargosa Desert, Nevada, with Special Reference to the Effects of Possible New Withdrawals of Ground Water*, Report TEI-840, U.S. Geological Survey, September 1963.

53. "Activities and Accomplishments of the Joint Committee on Atomic Energy During the 1st Session of the 88th Congress," Cong. Rec. (January 10, 1964), 265–266.

54. U.S. Department of Energy, *Atomic Power in Space: A History*, DOE/NE/32117-H1 (Washington, D.C.: U.S. Department of Energy, 1987), 39–40.

55. George Swink to George Hennan, December 11, 1965, permit application 19831, DWR-Carson City.

56. Associated Press, "Californian Says He'll Build City Near Nevada Test Site," *San Bernardino Sun*, February 20, 1964, C1.

57. George Swink to Tom Smales, March 5, 1962, permit application 19858, DWR-Carson City.

58. "Two Men Hit by Car, Killed," *Los Angeles Times*, February 1, 1963. Swink also referenced a "very bad accident" he was in "several years back" in a letter to George Hennan, December 11, 1965, permit application 19831, DWR-Carson City.

59. George Hennan to George Swink, December 8, 1965, permit application 19831, DWR-Carson City; and George Swink to Tom Smales, March 1, 1966, permit application 22984, DWR-Carson City. See also Dudley and Larson, *Effect of Irrigation Pumping on Desert Pupfish Habitats in Ash Meadows, Nye County, Nevada*, 16–18, for a list of wells and dates in Ash Meadows.

60. Lyle K. Gross, as told to Tamara Andreeva, "Land Grab," *Las Vegas Sun*, Sunday magazine section, November 4, 1962.

61. Permit application 20764, October 4, 1962, DWR-Carson City.

62. John Bradford, d. January 19, 1931, death certificate, Los Angeles County Registrar/Recorder-County Clerk.

63. After seeing the *Las Vegas Sun* article, Death Valley staff first contacted the Nevada Division of Water Resources for more information; see John A. Aubuchon to State Engineer, November 5, 1962, permit application 20764, DWR-Carson City.

64. Allison van V. Dunn to Regional Director, Western Region, NPS, December 14, 1962, box 1972, P11, RG 79, NARA-College Park.

65. G. F. Worts Jr., "Effect of Ground-Water Development on the Pool Level in Devils Hole, Death Valley National Monument, Nye County, Nevada" (Carson City, NV: U.S. Geological Survey, Water Resources Division, August 1963), 5.

66. E. W. Reed to Regional Director, Western Region, NPS, October 4, 1963, box 1972, P11, RG 79, NARA-College Park.

67. F. N. Dondero to John A. Aubuchon, November 14, 1962, box 1972, P11, RG 79, NARA-College Park. Correspondence also in permit application 20764 files at DWR-Carson City. In 1967, DWR denied the application on the grounds that it was "detrimental to the public interest." See Roland D. Westergard, Ruling in the matter of application 20764, October 11, 1967, permit application file 20764, DWR-Carson City.

68. James Cahalan, *Edward Abbey: A Life* (Tucson: University of Arizona Press, 2001), 101.

69. Robert R. Miller, "Status of Populations of Native Fishes of the Death Valley System in California and Nevada," NPS report, August 4, 1967, p. 2, folder "Natural Resources—Ichthyology," library, NPS-Cow Creek.

70. George Swink to Tom Smales, July 19, 1966, permit application 23279, DWR-Carson City.

71. Robert J. Bulkley Jr., *At Close Quarters: PT Boats in the Pacific* (Washington, D.C.: GPO, 1962), 230, 239–240, 500.

72. "Francis L. Cappaert, 'Love That Government Business,'" *Forbes*, September 1, 1966, 21–22.

73. John McAleenan, "'Silent Tycoon' Follows Path of Howard Hughes," *Enquirer and News* (Battle Creek, MI), December 16, 1973, A7.

74. Alvin A. Butkis, "The Silent Tycoon," *Dun's*, January 1972, 38.

75. Ibid., 40.

76. McAleenan, "'Silent Tycoon,' *Enquirer and News*, A7.

77. Butkis, "The Silent Tycoon," *Dun's*, 40.

78. The closest I have gotten to Cappaert's initial connection to Nevada and Ash Meadows is a Las Vegas development. In 1966, Guerdon Industries collaborated with Lockheed Aircraft and pioneering architect Paul Revere Williams to propose an aerial tram system for Las Vegas. Dorothy Wright, "When to Curve, When to Flow," *Desert Companion*, December 1, 2012, https://knpr.org/desert-companion/when-curve-when-flow.

79. McAleenan, "'Silent Tycoon' *Enquirer and News*, A7; and obituary, "F. L. Cappaert," *Clare (MI) Sentinel*, April 9, 1996, 14.

80. Lisa M. Fine, *The Story of Reo Joe: Work, Kin, and Community in Autotown, USA* (Philadelphia: Temple University Press, 2004), 149–168.

81. U.S. State Department cables 1973SANJO03194 (September 7, 1973), 1973STATE193374 (September 28, 1973), 1974SANJO04087 (October 24, 1974), and 1974STATE249145 (November 12, 1974), electronic telegrams, 1973–1974 (State Department Cables), RG 59, NARA-Online. Records searched through https://aad.archives.gov/aad/series-description.jsp?s=4073&cat=all&bc=sl on May 30, 2018.

82. Morton Mintz, "Winners Still Get Money," *Washington Post*, February 4, 1973, A1; and Morton Mintz, "Ten Donors Gave Nixon $6 Million," *Washington Post*, September 30, 1973, A1. On Cappaert's contributions, see also "420 Francis L. Cappaert," numerical files, Watergate Special Prosecution Force, Campaign Contribution Task Force, RG 460, NARA-Online. Cappaert was not charged with any crime in connection with his political contributions.

83. George Swink to Tom Smales, July 19, 1966, permit application 23279, DWR-Carson City.

84. Carol Hardy Vincent, Laura A. Hanson, and Carla N. Argueta, *Federal Land Ownership: Overview and Data*, Congressional Research Service Report R42346 (Washington, D.C.: Library of Congress, March 3, 2017).

85. Megan Black, *The Global Interior: Mineral Frontiers and American Power* (Cambridge, MA: Harvard University Press, 2018), 17.

86. See "A Proposed Classification of Public Lands in Nevada and Multiple Use Management in Nye County," Bureau of Land Management, Nevada Land Office, December 1968, box 98, Alan Bible papers, NC01, Special Collections, University Libraries, University of Nevada, Reno; and James Muhn and Hanson R. Stuart, *Opportunity and Challenge: The Story of* BLM (Washington, D.C.: GPO, 1988). On the broader history of the Bureau of Land Management, see also Skillen, *The Nation's Largest Landlord.*

87. Taylor Grazing Act of 1934, Section 8, Pub. L. No. 73-482, 48 Stat. 1269, at 1272.

88. Charles F. Wheatley Jr., *Study of Land Acquisitions and Exchanges Related to Retention and Management or Disposition of the Federal Public Lands*, prepared for the Public Land Law Review Commission (Springfield, VA: Federal Scientific and Technical Information Service, 1970), 52–53, 62–63.

89. Donald L. Barlett and James B. Steele, *Empire: The Life, Legend, and Madness of Howard Hughes* (New York: Norton, 1979), 173. Other land exchanges in Nevada also proved suspect during the postwar period. In 1960, the General Accounting Office found that the value of the lands BLM gave up in some large swaps in the state exceeded the value of lands it received in return. See General Accounting Office, Report B-141343, January 5, 1960, https://www.gao.gov/products/432328#mt=e-report.

90. "Federal-Private Land Exchange Shown in Records," *Nevada State Journal*, January 9, 1969, 2.

91. E. A. Moore to Dennis Hess, November 1, 1966, N-569, File 27-70-0038, box 138, patented land entry series case files, Nevada state office, NARA-San Bruno. Cappaert did not actually own this land to exchange with the government. His company, Springs Meadows Inc., held an option on private land in northern Nevada for the express purpose of using it to exchange with the federal government for Ash Meadows land. For a more detailed account of the BLM—Cappaert land exchange, see Kevin C. Brown, *Recovering the Devils Hole Pupfish: An Environmental History* (National Park Service, 2017), 72–103.

92. Mildred Johnson to John Hillsamer, October 16, 1967, N-569, file 27-70-0038, box 138, patented land entry series case files, Nevada state office, NARA-San Bruno.

93. Articles of Incorporation, Spring Meadows Inc., June 16, 1966, N-209, box 3, unpatented serialized land entry case files, Nevada State Office, NARA-San Bruno.

94. "Seven Major Committees Have New Chairmen," *Congressional Quarterly Weekly Report*, 27, no. 2 (January 10, 1969):68. Senate ethics rules with restrictions on outside employment went into effect in 1968. In 1977, providing professional services for compensation was banned with the committee report explicitly singling out the practice of law. See Jim Sasser, "Learning from the Past: The Senate Code of Conduct in Historical Perspective," *Cumberland Law Review* 8 (1977):357–384.

95. Alan Bible, senate campaign contributions and expenditures, 1956, 1962, Office of the Secretary of the Senate, Records of the U.S. Senate, RG 46, Center for Legislative Archives, NARA-online. Letterhead from 1970 shows Paul Bible listed as one of the firm's lawyers, see FWS-Las Vegas records for examples.

96. Paul Meiling, Bureau of Sport Fisheries and Wildlife telephone conversation record, with John Hillsamer, June 5, 1972, FWS-Sacramento.

97. Field Data for Spring Meadows applications, N-209, 210, 211, 212, 213, 376, 569, and 1033, April 26–June 5, 1967, N-209, box 3, unpatented serialized land entry case files, Nevada State Office, NARA-San Bruno.

98. "Notice of Proposed Classification," 32 Fed. Reg. 10377 (July 14, 1967).

99. John W. Stratton to Nolan F. Keil, August 29, 1967, N-209, box 3, unpatented serialized land entry case files, Nevada State Office, NARA-San Bruno.

100. E. W. Reed to Chief, "Office of Land and Water Rights, SSC," NPS, October 5, 1967, box 1972, P11, RG79, NARA-College Park.

101. Ibid.

102. Rolla E. Chandler to Dennis Hess, March 26, 1968, N-209, box 3, unpatented serialized land entry case files, Nevada State Office, NARA-San Bruno.

103. John W. Stratton to Dennis Hess, October 27, 1967, N-209, box 3, unpatented serialized land entry case files, Nevada State Office, NARA-San Bruno.

104. "Notice of Classification, May 6, 1968," 33 Fed. Reg. 7123–7124 (May 14, 1968).

105. Well Log and Report to the State Engineer of Nevada, Log 9312, December 8, 1966, accessed through online water permit information for water permit application 23057, Nevada Division of Water Resources. The well driller's report lists Swink's company (Nye County Land Company) as the owner of the well, but by this time it was owned by Cappaert's Spring Meadows Inc.

106. Dudley and Larson, *Effect of Irrigation Pumping*, 20.

107. Ed L. Reed, "A Study of Water Resources of Spring Meadows, Nye County, Nevada," p. 9, prepared for Spring Meadows Inc., Vicksburg, MS, June 1967, FWS-AMNWR. Reed wrote in the report that he acquired unpublished data at USGS offices, which is where he may have learned that the Devils Hole water level did not dip during the pump test.

108. Reed, "A Study of Water Resources," 10.

109. William Fields, "Trip Report," October 25–27, 1967, FWS-Ash Meadows.

Chapter 4. Save the Pupfish

1. Description based on Borovicka and Lostetter memos (n2–3) and color slides from Phil Pister personal papers.

2. Clinton Lostetter to John D. Findlay, May 28, 1969, FWS-Las Vegas; and William Fields, "Trip Report," October 25–27, 1967, FWS-Ash Meadows.

3. Robert Borovicka, "Special Report: Endemic Native Fishes of the Death Valley Drainage System," [May 1969], folder 002, series 001, pupfish activities, NPS-Cow Creek.

4. On the illegal well drilled closest to Devils Hole, see Right of Way application file N-660, BLM-Reno, especially Rolla Chandler to Spring Meadows Inc., July 2, 1969. On all wells drilled in Ash Meadows by Swink and Cappaert, see Dudley and Larson, *Effect of Irrigation Pumping on Desert Pupfish Habitats in Ash Meadows, Nye County, Nevada*, 16–18.

5. Dudley and Larson, *Effect of Irrigation Pumping*, 3.

6. In 1969, just as the issues facing Devils Hole came into view, BLM biologists had been planning two small reservations of public domain land in Ash Meadows around School and Jackrabbit Springs, home to Warm Springs pupfish, Ash Meadows Amargosa pupfish, and Ash Meadows speckled dace. Protecting these species and habitats became subsumed under the seemingly more critical threat facing Devils Hole and its pupfish. See Lewis H. Myers, "Management of Endangered Fish Habitat on Public Lands in the Ash Meadows Area, Nye County, Nevada," October 29, 1969, box 39, James D. Yoakum papers (2013-27), Special Collections, University Libraries, University of Nevada, Reno; and Lewis H. Myers, "Habitat Management Plan: Ash Meadows Pupfish Habitat," N5-WHA-AI, 1969, Bureau of Land Management, Las Vegas District, FWS-AMNWR.

7. Clinton Lostetter to John D. Findlay, May 28, 1969, FWS-Las Vegas. On this visit, see also recollection of one of the participants, Edwin Philip Pister, "The Desert Fishes Council: Catalyst for Change," in *Battle Against Extinction: Native Fish Management in the American West*, eds. W.L. Minckley and James E. Deacon (Tucson: University of Arizona Press, 1991), 57.

8. Edwin P. Pister, ed., *The Rare and Endangered Fishes of the Death Valley System—A Summary of the Proceedings of a Symposium Relating to their Protection and Preservation* [Proceedings of the Desert Fishes Council, vol. 1], (Bishop, CA: Desert Fishes Council, 1970), 8.

9. Anders Halverson, *An Entirely Synthetic Fish: How Rainbow Trout Beguiled America and Overran the World* (New Haven, CT: Yale University Press, 2010), 94–113.

10. 83 Fed. Reg. 38660, 38727–8, 38738 (August 7, 2018).

11. "Endangered Species," 32 Fed. Reg. 4001 (March 11, 1967).

12. David S. Wilcove and Margaret McMillan, "The Class of '67," in *The Endangered Species Act at Thirty*, vol. 1, *Renewing the Conservation Promise* (Washington: Island Press, 2006), 45–50; and Joe Roman, *Listed: Dispatches from America's Endangered Species Act* (Cambridge, MA: Harvard University Press, 2011). On current total threatened and endangered listings, see U.S. Fish and Wildlife Service, "Listed Species Summary (Boxscore)," https://ecos.fws.gov/ecpo/reports/box-score-report. As of October 11, 2018, it lists 1,662 U.S. species of plants and animals as threatened or endangered. (The list also contains a further 684 species that do not occur in the United States or its territory.)

13. Endangered Species Preservation Act, Section 1 (c), Pub. L. 89-669 (October 15, 1966).

14. Migratory Bird Treaty Act of 1918; and Bald Eagle Protection Act of 1940. On development of wildlife and endangered species laws in the U.S. during the 1960s, see Alagona, *After the Grizzly*, 100–101.

15. Endangered Species Preservation Act, Section 1 (b), Pub. L. 89-669 (October 15, 1966). The act also consolidated federal wildlife refuges and preserves into a unified "national wildlife refuge system" and authorized the usage of $15 million from Land and Water Conservation Fund Act monies for the acquisition of habitat for endangered species.

16. Barrow, *Nature's Ghosts*, 315–322.

17. On Miller's professional roles, see Cashner et al., "Robert Rush Miller (1916–2003)," 345. On his work in the BSFW list, see also Halverson, *An Entirely Synthetic Fish*, 108.

18. Committee on Rare and Endangered Wildlife Species, "Devils Hole pupfish," in *Rare and Endangered Fish and Wildlife of the United States*, Resource Publication 34 (Washington, D.C.: Bureau of Sport Fisheries and Wildlife, July 1966), F-24.

19. "Notice of Proposed Classification," 32 Fed. Reg. 10377 (July 14, 1967); and "Notice of Classification," 33 Fed. Reg. 7123 (May 14, 1968).

20. John W. Stratton to Dennis Hess, October 27, 1967, N-209, box 3, unpatented serialized land entry case files, Nevada State Office, NARA-San Bruno.

21. Associated Press, "78 Species Listed Near Extinction," *New York Times*, March 12, 1967, 46.

22. Clinton Lostetter to John D. Findlay, May 28, 1969, FWS-Las Vegas.

23. Pister, *The Rare and Endangered Fishes of the Death Valley System*, 1970, appendix pp. d–e.

24. In retirement, Robert J. Murphy, the Death Valley superintendent in 1969, wrote a book detailing the capture of the Manson family. See Bob Murphy, *Desert Shadows: A True Story of the Charles Manson Family in Death Valley* (Morongo Valley, CA: Sagebrush Press, 1993). On the Manson family and Death Valley, see also Vincent Bugliosi with Curt Gentry, *Helter Skelter: The True Story of the Manson Murders* (New York: Norton, 1974).

25. Pister, *The Rare and Endangered Fishes of the Death Valley System*, 1970, 7.

26. On changes within NPS during the 1960s and 1970s, see Sellars, *Preserving Nature in the National Parks*, 204–266.

27. Edwin P. Pister, "Wilderness Fish Stocking: History and Perspective," *Ecosystems* 4, no. 4 (2001):279–286.

28. Edwin Philip Pister, "Preserving Native Fishes and Their Ecosystems: A Pilgrim's Progress, 1950s–Present," an oral history conducted by Ann Lage in 2007–2008, Regional Oral History Office, The Bancroft Library, University of California, Berkeley, 2009, 96.

29. Ibid., 96.

30. Edwin P. Pister to Robert Rush Miller, n.d. (ca. 1990–1991), folder "Pister," biographic files, Miller papers, UMMZ.

31. Pister, "Preserving Native Fishes and Their Ecosystems," 96.

32. Edwin P. Pister, "Species in a Bucket," *Natural History* 102, no. 1 (1993):14–18.

33. Pister, *The Rare and Endangered Fishes of the Death Valley System*, 1970, 2.

34. Endangered Species Conservation Act, Pub. L. 91-135 (December 5, 1969).

35. The lone Washington attendee was an old friend of Carl Hubbs, NPS's chief aquatic biologist, Orthello L. Wallis. Wallis had earlier written an agency report about Death Valley's fishes. See O. L. Wallis, "Interpretation and Protection of Desert Fishes of Death Valley National Monument," April 1959, folder "Natural Resources—Ichthyology," library, NPS-Cow Creek.

36. Pister, *The Rare and Endangered Fishes of the Death Valley System*, 1970, 13.

37. The Devils Hole pupfish is not the only endangered species that struggled to attract attention from agency leaders. The first meetings that led to the recovery of the Florida panther were equally modest. See Craig Pittman, *Cat Tale: The Wild, Weird Battle to Save the Florida Panther* (Toronto: Hanover Square Press, 2020), 57–72.

38. Robert Miller and NPS personnel seem to have played key roles in communicating concern over the situation in Ash Meadows to Interior Department leadership. See Miller to Walter J. Hickel (Secretary of the Interior), January 6, 1970, and Edward Hummel (Assistant Director, NPS) to Miller, February 5, 1970, both folder "Ash Meadows Project," Miller papers, UMMZ.

39. Peter Sanchez to Robert R. Miller, February 16, 1970, series 002, pupfish activities, NPS-Cow Creek.

40. Leslie Glasgow to Charles Meacham, February 16, 1970, box 2757, P11, RG 79, NARA-College Park.

41. John Gottschalk to Leslie Glasgow, March 5, 1970, box 2757, P11, RG 79, NARA-College Park.

42. John Gottschalk to Leslie Glasgow, March 5, 1970, box 2757, P11, RG 79, NARA-College Park. For Miller original, see Robert R. Miller, "Symposium on Rare and Endangered Fishes of the Death Valley System: Summary Report to the National Park Service," December 7, 1969, FWS-Las Vegas.

43. John P. Linduska to Harry Goodwin, April 3, 1970, FWS-Las Vegas. On BSFW employees concerned about Gottschalk's response to Goodwin, see also Travis Roberts to Gottschalk, April 1, 1970, FWS-Las Vegas. NPS also remained frustrated with the lack of movement in the Interior Department. See Edward A. Hummel to Leslie Glasgow, April 1, 1970, box 2757, P11, RG 79, NARA-College Park.

44. Robert Murphy to Director, NPS, May 6, 1970, FWS-Las Vegas.

45. On National Park Service efforts to keep Devils Hole off of maps in the 1950s, see ch. 2. On scientists' attitudes toward publicizing Devils Hole, Robert R. Miller to Carl Hubbs, March 23, 1964; and Martin Brittan, "Some Preliminary Observations Concerning Death Valley Area Pupfishes," ca. February 1964, 1–2, both in Miller papers, UMMZ.

46. Pister, *The Rare and Endangered Fishes of the Death Valley System*, 1970, 18.

47. Publicity that the agencies were acutely aware of is recounted in O. L. Wallis and E. W. Reed to Assistant Director, Park Management, NPS, May 7, 1970, box 2757, P11, RG 79, NARA-College Park. See also *Cry California: The Journal of California Tomorrow* 5, no. 2 (Spring 1970): 1–25; and *Timetable for Disaster*, documentary, 1970. Examples of letters from the public can be found in box 2757, P11, RG 79, NARA-College Park. Also important for establishing the task force was a conversation between Commissioner of Wildlife Charles Meacham and Phil Pister, early May 1970; see Pister, "The Desert Fishes Council: Catalyst for Change," in *Battle against Extinction*, 60.

48. O. L. Wallis and E. W. Reed to Assistant Director, Park Management, NPS, May 7, 1970, box 2757, P11, RG 79, NARA-College Park.

49. James McBroom, "Pupfish Task Force Report," in *Proceedings of the Desert Fishes Council*, ed. Edwin P. Pister, vol. 3, 1971 (December 1983), appendix 1971-5, 22.

50. Reported in Clinton Lostetter, "Summary Interagency Field Task Force Devils Hole Pupfish Meeting at Furnace Creek, Death Valley, California, May 25, 1970," 1, FWS-Sacramento.

51. Louis Boll to James T. McBroom, August 24, 1970, FWS-Las Vegas.

52. Dudley and Larson, *Effect of Irrigation Pumping*, 1, 3.

53. U.S. Department of the Interior, "A Task Force Report on Let's Save the Desert Pupfish," Washington, D.C., June 1970, 6, FWS-Las Vegas.

54. Peter Sanchez, "Inter-Agency Field Task Force Meeting, Devils Hole Pupfish, Death Valley National Monument, May 25, 1970," FWS-Las Vegas.

55. Tina Nappe to Richard G. Miller, "Devils Hole Interagency Meeting," ca. June 1970, folder 17, box 14, Foresta records, UNR.

56. Tina Nappe to James McBroom, July 17, 1970, FWS-Las Vegas.

57. Harrison Loesch to Director, BLM, September 3, 1970, N-209, box 3, unpatented serialized land entry case files, Nevada State Office, NARA-San Bruno.

58. Foresta Institute, "Statement on the Pupfish," ca. November 1970, folder 20, box 14, Foresta records, UNR.

59. Pister, *The Rare and Endangered Fishes of the Death Valley System*, 1970, 8. On the role of science in shaping conservation at Devils Hole, see also Brown, "'An Exceedingly Simple, Little Ecosystem.'"

60. Carol Jeanne James, "Aspects of the Ecology of the Devils Hole Pupfish, *Cyprinodon diabolis* Wales" (master's thesis, University of Nevada, Las Vegas, 1969).

61. Carol J. James, field notes, June 30 and July 1, 1967, Carol J. James field notes collection, MVZA.MSS.0458, Museum of Vertebrate Zoology, University of California, Berkeley.

62. Curt Meine, Michael Soulé, and Reed F. Noss, "'A Mission-Driven Discipline': The Growth of Conservation Biology," *Conservation Biology* 20, no. 3 (2006):631–651; Timothy J. Farnham, *Saving Nature's Legacy: Origins of the Idea of Biological Diversity* (New Haven, CT: Yale University Press, 2007); and Michael E. Soulé and Bruce A. Wilcox, eds., *Conservation Biology: An Evolutionary-Ecological Perspective* (Sunderland, MA: Sinauer Associates, 1980).

63. Barrow, *Nature's Ghosts*, 301–306.

64. Contemporaneous notes made during Pister's presentation, Desert Fishes Council, Furnace Creek, CA, November 15, 2018, in author's possession.

65. Peter G. Sanchez to Regional Coordinator, Environmental Early Warning System, April 29, 1970, series 002, pupfish activities, NPS-Cow Creek.

66. Peter G. Sanchez to Robert Murphy, May 25, 1970, FWS-Las Vegas.

67. E. W. Reed to James McBroom, July 22, 1970, FWS-Las Vegas.

68. Lawrence Hadley to Director, Western Region, NPS, December 10, 1970, box 2757, P11, RG 79, NARA-College Park.

69. Robert Miller to Phil Pister, November 22, 1970, FWS-Las Vegas; Dale Lockard to Robert R. Miller, December 24, 1970, Edwin P. Pister personal papers (hereafter Pister papers); and Dale Lockard to Director, NDFG, January 15, 1971, NDOW-Las Vegas.

70. O. L. Wallis to James McBroom, August 11, 1970, FWS-Las Vegas.

71. Robert Miller to Phil Pister, March 17, 1972, Pister papers.

72. Robert Miller to Robert Murphy, April 10, 1972, FWS-Las Vegas.

73. Robert R. Miller to Peter Sanchez, May 28, 1970, Sanchez Files, natural resources collection, NPS-Cow Creek.

74. Robert R. Miller to Phil Pister, April 11, 1972, FWS-Las Vegas.

75. Robert R. Miller to Robert Linn, April 15, 1972, series 004, pupfish activities, NPS-Cow Creek.

76. John Findlay to James Thompson, June 15, 1973, FWS-Las Vegas.

77. Phil Pister to W. Stephenson, February 8, 1974, FWS-Las Vegas.

78. On lights: Robert Miller to James Thompson, January 15, 1973, FWS-Las Vegas. On shelf removal: Dean Garrett to "Chief I&RM," December 19, 1974, series 006, pupfish activities, NPS-Cow Creek.

79. Robert Rush Miller, "Preliminary List of Rare, Restricted and/or Threatened North American Freshwater Fishes," American Society of Ichthyology and Herpetology, May 1963, 7, folder 13, box 51, Hubbs papers.

80. Miller, "Status of Populations of Native Fishes of the Death Valley System in California and Nevada," Completion Report of Resource Studies Problem Undertaken for the U.S. National Park Service, June-July 1967, 10, folder "Natural Resources—Ichthyology," library, NPS-Cow Creek.

81. Howard W. Baker to Regional Director, Western Region, NPS, August 25, 1967, N1423, box 2300, P11, RG 79, NARA-College Park.

82. Phil Pister to William F. Richardson, August 25, 1970, FWS-Las Vegas.

83. Dale Lockard, "Transplant of *Cyprinodon diabolis* from Devils Hole to Point of Rocks Spring," September 12, 1970, NDOW-Las Vegas.

84. Dale Lockard to James McBroom, September 17, 1970, NDOW-Las Vegas.

85. Peter G. Sanchez to Robert Murphy, April 20, 1972, series 020, pupfish activities, NPS-Cow Creek; and Grapevine Subdistrict Ranger, "Log of Events at Grapevine Springs Devils Hole Pupfish Refuge," October 15, 1972, series 020, pupfish activities, NPS-Cow Creek.

86. Dale Lockard, "Devils Hole Transplant Field Notes," June 21, 1972, NDOW-Las Vegas; and Charles Osborn, "Monthly Status Report—August thru October 4, 1972—Endangered Desert Fishes," FWS-Las Vegas.

87. James McBroom to task force, January 28, 1971, Pister papers; and O. L. Wallis, "Notes by O. L. Wallis," ca. January 1971, series 003, pupfish activities, NPS-Cow Creek.

88. Robert R. Miller to James McBroom, July 7, 1971, Pister papers; and Charles Osborn, field notes, November 30, 1971, series 003, pupfish activities, NPS-Cow Creek.

89. Pister, *Proceedings of the Desert Fishes Council*, 1972, vol. 4, 1972 (December 1983), 57; and Charles Osborn, "Status Report—October 4, 1972 thru Jan. 15, 1973—Endangered Desert Fishes," FWS-Las Vegas.

90. Charles E. Osborn, "Status Report—October 4, 1972 thru Jan. 15, 1973," FWS-Las Vegas.

91. Robert R. Miller to Carl Hubbs, May 18, 1937, folder 11, box 230, Hubbs papers.

92. Evidence for other aquarium attempts before 1970 come from Gilbert Curtright to Carl Hubbs, January 6, 1950, folder 4, box 239, Hubbs papers; Robert K. Liu to Louis Garibaldi, January 5, 1971, NPS-Pahrump.

93. Miller, "Status of Populations," 10.

94. Alan McCready to James Deacon, July 12, 1970, Chuck Douglas files, resource management collection, NPS-Cow Creek; and Alan McCready to James Deacon, March 27. 1971, series 003, pupfish activities, NPS-Cow Creek.

95. Dale Lockard, narrative report on the collection and shipment of Devils Hole pupfish to Steinhart Aquarium, NDOW-Las Vegas; "Steinhart Aquarium Log for Devils Hole Pupfish, *Cyprinodon diabolis*," October 5, 1971 and November 4, 1971, FWS-Las Vegas; and "Steinhart Aquarium Log for Devils Hole pupfish, *Cyprinodon diabolis*," July 27, 1972, Pister papers.

96. Richard Haas, reports no. 1 (September 1972) and no. 2 (January 1973), series 021, pupfish activities, NPS-Cow Creek.

97. Dale Lockard to Harold Thompson, October 4, 1972, NDOW-Las Vegas.

98. U.S. Bureau of Reclamation, *Interim Report on Inland Basins Projects, Nevada-California, Amargosa Project: Reconnaissance Investigation*, Bureau of Reclamation, Region 3, Boulder City, Nevada, July 1968; and U.S. Bureau of Reclamation, *Amargosa Project: California-Nevada, Concluding Report*, Boulder City, NV: U.S. Bureau of Reclamation, February 1975.

99. On dimensions of the Hoover Dam refuge, see J. E. Williams, *Observations on the Status of the Devils Hole Pupfish in the Hoover Dam Refugium*, Report REC-ERC-77-11, (Denver, CO: U.S. Bureau of Reclamation, September 1977), 3.

100. Charles E. Osborn, "Status Report—October 4, 1972 thru January 15, 1973," FWS-Las Vegas; and Williams, *Observations on the Status of the Devils Hole Pupfish in the Hoover Dam Refugium*, 1.

101. Robert R. Miller to James Thompson, January 15, 1973, FWS-Las Vegas.

102. Dean Garrett to Chief Ranger, July 13, 1973, series 005, pupfish activities, NPS-Cow Creek.

103. Thomas M. Baugh and James E. Deacon, "Evaluation of the Role of Refugia in Conservation Efforts for the Devils Hole Pupfish, *Cyprinodon diabolis* Wales," *Zoo Biology* 7 (1988):355.

104. For full accounting of these, see Brown, *Recovering the Devils Hole Pupfish*, 139–140, 163 (tables 3.1, 3.3).

105. Dale Lockard to Phil Pister, February 17, 1971, Pister papers.

Chapter 5. Kill the Pupfish

1. See, for example, Peter G. Sanchez to Robert Murphy, May 25, 1970, FWS-Las Vegas; and Dale V. Lockard, "Transplant of *Cyprinodon diabolis* from Devils Hole to Point of Rocks Spring," September 12, 1970, NDOW-Las Vegas.

2. An act providing for the protection and propagation of native fauna and flora, ch. 446, 1969 Nev. Stat. 773; and Leona K. Svancara et al., "Endangered Species Time Line," in *The Endangered Species Act at Thirty*, vol. 2, *Biodiversity in Human-Dominated Landscapes*, eds. J. Michael Scott, Dale D. Goble, and Frank W. Davis (Washington, D.C.: Island Press, 2006), 26.

3. *The Journal of the Assembly of the Fifty-Fifth Session of the Legislature of the State of Nevada* (State Printing Office: Carson City, Nevada), 666; and *The Journal of the Senate of the Fifty-Fifth Session of the Legislature of the State of Nevada* (Carson City, NV: State Printing Office), 747.

4. Dale V. Lockard, "Nevada Fish and Game," *Las Vegas Sun*, December 12, 1969; and Charles Crunden, "Plenty of Trout Stocked in Lake Mead," *Las Vegas Sun*, December 12, 1969.

5. M. Kent (Tim) Hafen to Nevada Department of Fish and Game, July 26, 1973, NDOW-Las Vegas.

6. Dale V. Lockard to Phil Pister, n.d., handwritten on copy of Hafen to Nevada Department of Fish and Game, July 26, 1973 letter, Pister papers.

7. A. D. Hopkins, "Anti-Pupfish Movement Underway," *Las Vegas Sun*, August 10, 1973. The article misspelled Bob Ruud's last name as "Rudd."

8. "Save the Pupfish" sticker can be seen in photograph accompanying Lynn Lilliston, "Gallant Saga of Pupfish's Survival," *Los Angeles Times*, November 29, 1970.

9. See chapter 3, this book; and Ed L. Reed, "A Study of Water Resources of Spring Meadows, Nye County, Nevada," prepared for Spring Meadows Inc., Vicksburg, MS, June 1967, 9, FWS-Ash Meadows.

10. Paul Sabin, "Environmental Law and the End of the New Deal Order," *Law and History Review* 33, no. 4 (2015):965–1003. On shifting environmental politics, policy, and law in the early 1970s, see also Adam Rome, *The Genius of Earth Day: How a 1970 Teach-In Unexpectedly Made the First Green Generation* (New York: Hill and Wang, 2013); Richard N. L. Andrews, *Managing*

the Environment, Managing Ourselves: A History of American Environmental Policy, 2nd ed. (New Haven, CT: Yale University Press, 2006); and James Morton Turner, *The Promise of Wilderness: American Environmental Politics since 1964* (Seattle: University of Washington Press, 2012).

11. Endangered Species Act of 1973, Pub. L. No. 93-205, 87 Stat. 884 (December 28, 1973) at 900.

12. İsmail K. Sağlam et al., "Phylogenetics Support an Ancient Common Origin of Two Scientific Icons: Devils Hole and Devils Hole Pupfish," *Molecular Ecology* 25 (2016):3962; doi:10.1111/mec.13732.

13. Robert R. Miller to Joel W. Hedgpeth, April 17, 1970, Miller papers, UMMZ.

14. The term for a suit to compel the government to act on its existing administrative authority is called a "mandamus action." H. Donald Harris Jr. to Leslie Glasgow, May 6, 1970, FWS-Las Vegas.

15. Charles H. Vaughn to Associate Director, BSFW, April 22, 1970, FWS-Las Vegas.

16. Both earthquakes and groundwater pumping on the Nevada Test Site do influence the water level in Devils Hole, though less dramatically than the onset of Cappaert's pumping in Ash Meadows in the late 1960s. See Halford and Jackson, *Groundwater Characterization and Effects of Pumping in the Death Valley*, 154.

17. U.S. Geological Survey, "Summary of Known Ground Water Conditions in the Vicinity of Devils Hole, Southwestern Nevada, and Recommendations for Additional Investigations," May 15, 1970, FWS-Las Vegas. On 1950s and 1960s vandalism at Devils Hole, see Brown, *Recovering the Devils Hole Pupfish*, appendix C, 653–654.

18. Charles H. Vaughn to Associate Director, BSFW, April 22, 1970, FWS-Las Vegas.

19. Harthon L. Bill to Neil Humphrey, May 20, 1970, box 2757, P11, RG 79, NARA-College Park.

20. BSFW, "Status of the Pupfish Project," May 20, 1970, FWS-Las Vegas.

21. G. William Fiero and G. B. Maxey, *Hydrogeology of the Devils Hole Area, Ash Meadows, Nevada* (Reno: Center for Water Resources Research, Desert Research Institute, University of Nevada System, June 1970), https://archive.org/details/hydrogeologyofdeoofier/page/n61/mode/2up.

22. David A. Watts to James T. McBroom, June 19, 1970, series 023, pupfish activities, NPS-Cow Creek.

23. The budget office at Interior balked at the price tag for the study, forcing five department agencies involved in the task force to share the cost from reserve funds. See Brown, *Recovering the Devils Hole Pupfish*, 206–207.

24. Leslie Glasgow to Fred J. Russell, June 19, 1970, FWS-Las Vegas.

25. William Dudley, testimony from United States v. Cappaert, District Court hearing, July 3, 1972, transcript, 10, box 1, desert pupfish collection (MS 20), Special Collections, University of Nevada, Las Vegas (hereafter, desert pupfish collection, UNLV).

26. W. W. Dudley, "Status and Data Report, May 30, 1971," June 14, 1971, FWS-Las Vegas.

27. W. W. Dudley to Gerald Meyer, June 15, 1971, FWS-Las Vegas.

28. John D. Findlay to Acting Director, BSFW, March 23, 1972, FWS-Sacramento. Before the 1990s, Region 1 of the BSFW/FWS included Nevada and California.

29. Alan Bible to M. L. Barry, June 27, 1972, folder S. 2141, box 175, Alan Bible papers, NC01, Special Collections, University Libraries, University of Nevada, Reno. On Alan Bible's career and politics, see Gary E. Elliot, *Senator Alan Bible and the Politics of the New West* (Reno: University of Nevada Press, 1994).

30. On Cranston's bills for a "Desert Pupfish National Monument" and the BSFW proposal for an "Amargosa National Wildlife Refuge" in the early 1970s, see Brown, *Recovering the Devils*

Hole Pupfish, 253–300. On the tensions in the goals of the national wildlife refuge system, see Michael J. Bean, "Biological Diversity and the Refuge System: Beyond the Endangered Species Act in Fish and Wildlife Management," *Transactions of the Fifty-Seventh North American Wildlife and Natural Resources Conference*, ed. Richard E. McCabe (Washington, D.C.: Wildlife Management Institute, 1992), 577–580.

31. Based on analysis of Nevada Division of Water Resources water permit applications database (http://water.nv.gov/waterrights/), on April 21, 1970, Spring Meadows Inc. submitted sixteen groundwater permit applications (numbers 25551–25566) on the four public land survey system townships immediately surrounding Devils Hole (T17S R50E, T17S R51E, T18S R50E, T18S R51E Mount Diablo Meridian). In May and June 1970, it submitted additional groundwater permit applications (25579–25582, 25662).

32. Testimony of B. L. Barnett, transcript of December 16, 1970, water hearing before state engineer, 22, 27, FWS-Sacramento. Barnett estimated that a total of fourteen to fifteen wells had been drilled that "were intended to be producers," but six to seven did not do well on pump testing and were not in use (p. 29).

33. NPS submitted an official protest on August 21, 1970, for the following applications: 25556–25559, 25561–25564, 25571, 25581, and 25662. Protested applications listed on transcript of December 16, 1970, water hearing before state engineer, 7, FWS-Sacramento.

34. James McBroom to Roland Westergard, September 4, 1970, FWS-Las Vegas.

35. James McBroom to Roland Westergard, November 2, 1970, FWS-Las Vegas.

36. Roland Westergard to James McBroom, October 21, 1970, FWS-Las Vegas. Westergard referred McBroom to Nevada Rev. Stat. §533.370, "Approval or rejection of application by State Engineer: Conditions; exceptions; considerations; procedure."

37. Transcript of December 16, 1970, water hearing before state engineer, 67, FWS-Sacramento.

38. Roland Westergard, oral decision, December 16, 1970, 1–4, FWS-Las Vegas. Even if there had been an existing water right at Devils Hole, Westergard pointed out that a "reasonable" lowering of the water table was to be expected by existing water rights holders.

39. Roland Westergard, oral decision, December 16, 1970, 5, FWS-Las Vegas.

40. Otto Aho to David A. Watts, December 23, 1970, FWS-Las Vegas.

41. Winters v. United States, 207 U.S. 564 (1908). On the history of the *Winters* case and decision, see Norris Hundley Jr., "The 'Winters' Decision and Indian Water Rights: A Mystery Reexamined," *Western Historical Quarterly* 13, no. 1 (1982):17–42; John Shurts, *Indian Reserved Water Rights: The Winters Doctrine in its Social and Legal Context, 1880s-1930s* (Norman: University of Oklahoma Press, 2000); and Michael A. Massie, "The Defeat of Assimilation and the Rise of Colonialism on the Fort Belknap Reservation, 1875–1925," *American Indian Culture and Research* 7, no. 4 (1984):33–49.

42. Arizona v. California et al., 373 U.S. at 601 (1963). For context on *Arizona v. California*, see Hundley, *The Great Thirst*, 304–308, 533–534. The U.S. Supreme Court affirmed its argument that the federal government had the authority to reserve water rights for "any federal enclave" in 1971. See United States v. District Court of Eagle County, 401 U.S. 520 at 522-3 (1971).

43. Mitchell Melich to John Mitchell, July 22, 1971, box 2743, P11, RG 79, NARA-College Park. In addition to Melich's focus on Devils Hole, his office also proposed that the federal government pursue protection of springs on public land in Ash Meadows by invoking a 1926 executive order (Public Water Reserve No. 107) that reserved springs for public use, especially for the grazing of cattle. Pumping by Spring Meadows Inc. also affected spring flow across Ash Meadows and

might impede the use of springs for public land cattle grazing. The Department of Justice chose not to pursue this line of reasoning.

44. "Complaint Seeking Injunctive Relief," *United States v. Spring Meadows, Inc.* (Civil LV-1687), August 18, 1971, box 2, desert pupfish collection, UNLV.

45. "Stipulation for Continuance," *United States v. Spring Meadows, Inc.* (Civil LV-1687), September 2, 1971, box 2, desert pupfish collection, UNLV; and David A. Watts to James McBroom, September 1, 1971, FWS-Las Vegas.

46. Department of the Interior, news release, September 2, 1971, FWS-Las Vegas. The department did receive some news coverage of the case, including from the *Washington Post* (September 3, 1971) and the *Wall Street Journal* (September 7, 1971).

47. William Dudley to Gerald Meyer, December 10, 1971, series 004, pupfish activities, NPS-Cow Creek.

48. Chuck Osborn to John Findlay, January 26, 1972, series 004, pupfish activities, NPS-Cow Creek; and Cappaert Enterprises water permit application 26446, December 21, 1971, DWR-Carson City.

49. Ron Lambertson, quoted in Don Stewart memo for the files, February 18, 1972, FWS-Sacramento.

50. John D. Findlay to Acting Director, BSFW, March 3, 1972, FWS-Las Vegas; Daniel J. Tobin to Director, NPS, March 3, 1972, Pister papers; and Gerald Meyer to David Watts, February 24, 1972, Peter G. Sanchez Files, resource management collection, NPS-Cow Creek.

51. David R. Warner (for Kent Frizzell, Assistant Attorney General) to Mitchell Melich, May 19, 1972, FWS-Ash Meadows.

52. Bernard R. Meyer to Nathaniel Reed, June 5, 1972, series 023, pupfish activities, NPS-Cow Creek.

53. Roger D. Foley, "Finding of Fact, Conclusion of Law and Preliminary Injunction," June 5, 1973, *United States v. Cappaert*, Civil LV-1687, 0668, box 2, desert pupfish collection, UNLV.

54. Roger D. Foley, "Finding of Fact, Conclusion of Law and Preliminary Injunction," June 5, 1973, *United States v. Cappaert*, Civil LV-1687, 0671, box 2, desert pupfish collection, UNLV.

55. Samuel S. Lionel, "Memorandum of Points and Authorities in Opposition to Plaintiff's Amended Motion for Preliminary Injunction," June 1972, *United States v. Cappaert*, Civil LV-1687, 0135-0175, box 2, desert pupfish collection, UNLV.

56. Robert List and L. William Paul, "Motion to Intervene as a Defendant," June 30, 1972, *United States v. Cappaert*, Civil LV-1687, 0122–0123, box 2, desert pupfish collection, UNLV; and Roland Westergard to Robert List, June 21, 1972, and Roland Westergard to Mike O'Callaghan, June 21, 1972, permit file 25556, DWR-Carson City. On Nevada's legal reasoning, see Peter D. Laxalt and George Allison, "Intervenor, State of Nevada's Post-Hearing Brief," September 23, 1972, *United States v. Cappaert*, Civil LV-1687, 0325-0352, box 2, desert pupfish collection, UNLV.

57. Samuel S. Lionel, "Memorandum of Points and Authorities in Opposition to Plaintiff's Amended Motion for Preliminary Injunction," June 1972, *United States v. Cappaert*, Civil LV-1687, 0166-174, box 2, desert pupfish collection, UNLV; and United States v. Cappaert, 508 F.2d 313 (1974).

58. Roger D. Foley, "Minute Order," April 18, 1973, *United States v. Cappaert*, Civil LV-1687, 0585, box 2, desert pupfish collection, UNLV; and Roger D. Foley, "Finding of Fact, Conclusion of Law and Preliminary Injunction," June 5, 1973, *United States v. Cappaert*, Civil LV-1687, 0665, box 2, desert pupfish collection, UNLV.

59. Roger D. Foley, "Preliminary Injunction and Order Appointing Special Master," June 5, 1973, *United States v. Cappaert*, Civil LV-1687, 0675-0677, box 2, desert pupfish collection, UNLV.

60. Jack Anderson, "Administration Saves Desert Pupfish," *Washington Post*, April 27, 1973.

61. George Swink to Roland Westergard, September 27, 1973, Amargosa Drainage Basin–Ash Meadows correspondence, DWR-Carson City.

62. United States v. Cappaert, 508 F.2d 313 (9th Cir. 1974).

63. "Summary of the Meetings 1976," *Copeia* 1976, no. 4 (1976):848–858.

64. Cappaert v. United States, 426 U.S. 128 (1976) at 147.

65. Edwin Philip Pister, oral history interview by Ann Lage, 2007–2008, Regional Oral History Office, Bancroft Library, University of California, Berkeley, 139.

66. Phil Pister to Carl Hubbs and Robert Miller, June 7, 1976, handwritten note on University of Alaska memo pad; and Carl Hubbs to Phil Pister, June 7, 1976, folder 4, box 56, Hubbs papers, UCSD.

67. Cappaert v. United States, 426 U.S. 128 (1976) at 141. It was not impossible to frame the Supreme Court's ruling as an important limit on federal reserved rights. The federal government, Burger's opinion made clear, was not entitled to a full restoration of the Devils Hole pool level. Instead, it could only have that water "necessary" to support the pupfish, a level that, as described below, was substantially lower than before pumping began.

68. Cappaert v. United States, 426 U.S. 128 (1976) at 133.

69. Dick Rayner to Peter Sanchez and James Deacon, February 26, 1976, series 027, pupfish activities, NPS-Cow Creek. See also Andrew Walch to N. Gregory Austin, June 3, 1976, series 023, pupfish activities, NPS-Cow Creek.

70. Andrew Walch to N. Gregory Austin, June 22, 1976, Series 023, pupfish activities, NPS-Cow Creek; and Peter G. Sanchez to James Deacon, July 22, 1976, NDOW-Las Vegas.

71. Andrew Walch to N. Gregory Austin, November 3, 1976, series 023, pupfish activities, NPS-Cow Creek.

72. Notice of Motion and Motion to Modify Permanent Injunction (LV-1687), November 30, 1976, box 3, desert pupfish collection, UNLV. See also "Plaintiff's Post Hearing Brief in Support of Its Motion to Modify Injunction," June 17, 1977, box 3, desert pupfish collection, UNLV.

73. James Deacon, field notes, February 4, 1961, James Deacon personal papers, Henderson, Nevada, copy in author's possession. In 1961, UNLV was called the University of Nevada, Southern Division.

74. James Deacon, interview by Kevin C. Brown, April 3, 2014, Henderson, Nevada, transcript in author's possession, 2.

75. On Deacon's life, see Henry Brean, "Noted UNLV Biologist James Deacon Has Died," *Las Vegas Review-Journal*, February 27, 2015, https://www.reviewjournal.com/news/noted-unlv-biologist-james-deacon-has-died/; and Cindy Deacon Williams, "James E. Deacon," in *Standing between Life and Extinction*, eds. Propst et al., 22–26.

76. "Memorandum in Support of Motion to Modify Permanent Injunction," November 30, 1976, 388, box 3, desert pupfish collection, UNLV.

77. Defendant's Answering Brief, August 18, 1977, box 3, desert pupfish collection, UNLV.

78. Judge Roger Foley, Order, December 22, 1977, box 3, desert pupfish collection, UNLV. The case finally concluded on March 24, 1978, when Foley issued an order modifying the final injunction, confirming the 2.7-foot level as the extent of the federal reserved water right at Devils Hole. On the role of monitoring in defining the water right at Devils Hole, see Owen R. Williams et al., "Water Rights and Devils Hole Pupfish at Death Valley National Monument," in *Science and Ecosystem Management in the National Parks*, eds. William L. Halverson and

Gary E. Davis (Tucson: University of Arizona Press, 1996), 161–183; and Kevin P. Wilson, Mark B. Hausner, and Kevin C. Brown, "The Devils Hole Pupfish: Science in a Time of Crises," in *Standing between Life and Extinction*, eds. Propst et al., 379–389.

79. "17 States Lose Water Decision," *Nevada West and Pahrump Valley Times*, July 1976.

80. "Rottenest Thing I've Ever Seen," *Nevada West and Pahrump Valley Times*, July 1976.

81. William J. Newman to Roland Westergard, December 1, 1972, folder "Amargosa Drainage Basin—Ash Meadows Symposium," DWR-Carson City.

82. Samuel Lionel, oral argument, Cappaert v. United States, Supreme Court of the United States, January 12, 1976.

83. Roland D. Westergard "The Pupfish Caper," n.d., folder "Amargosa Basin, Ash Meadows, Endangered Species, Pupfish Correspondence," DWR-Carson City. Westergard may have delivered several versions of the Pupfish Caper speech during the 1970s. The comments by Westergard that apparently motivated Steninger's editorial came from an appearance by Westergard in early 1976 before the State Multiple Use Advisory Committee on Federal Lands. See Mike O'Callaghan to James E. Deacon, March 31, 1976, NPS-Pahrump.

84. Editorial, "The 'Pupfish Caper' in Proper Perspective," *Elko Daily Free Press*, March 8, 1976.

85. Joseph Wales to Phil Pister, March 15, 1976, Pister papers.

86. William H. Meyer, letter to the editor, *Elko Daily Free Press*, March 18, 1976.

87. Phil Pister, letter to the editor, "Pupfish and the Public Trust," *Inyo Register*, March 25, 1976, 26; and "Over the Line: Nevada News," *Inyo Register*, March 18, 1976, 28.

88. Editorial, "Rebuttal Arguments on 'Pupfish Caper,'" *Elko Daily Free Press*, March 22, 1976.

89. Editorial, "Ruling on Pupfish Predictable, Wrong," *Elko Daily Free Press*, June 8, 1976.

90. Molly Ivins, "Nevada Presses Its Effort to Regain Disputed Land," *New York Times*, July 16, 1979, A14.

91. R. McGreggor Cawley, *Federal Land, Western Anger: The Sagebrush Rebellion and Environmental Politics* (Lawrence: University Press of Kansas, 1993); Skillen, *The Nation's Largest Landlord*, 120–123; and Jacqueline Vaughn Switzer, *Green Backlash: The History and Politics of Environmental Opposition in the U.S.* (Boulder, CO: Lynne Rienner Publishers, 1997), 171–190.

92. Wayne Westphal, case incident record, July 23, 1977, series 009, pupfish activities, NPS-Cow Creek; Dan Dellinges, case incident report, March 23, 1979, series 011, pupfish activities, NPS-Cow Creek; and James Deacon, "Report on Devils Hole, May, June, and July 1981," and "Report on Devils Hole: August 1981," NPS-Pahrump. These are only some examples uncovered in the course of my research. For additional instances of illegal entry and vandalism at Devils Hole, see Brown, *Recovering the Devils Hole Pupfish*, appendix C, 652–661.

93. Veronica Rocha, "Guns, Beer, and Vomit: Rampage Leaves Endangered Fish Dead in Death Valley," *Los Angeles Times*, May 9, 2016, https://www.latimes.com/local/lanow/la-me-ln-reward-offered-death-valley-20160509-story.html.

94. Paige Blankenbuehler, "How a Tiny Endangered Species Put a Man in Prison," *High Country News*, April 15, 2019, https://www.hcn.org/issues/51.6/endangered-species-how-a-tiny-endangered-species-put-a-man-in-prison.

95. Kenneth M. Murchison, *The Snail Darter Case:* TVA *versus the Endangered Species Act* (Lawrence: University Press of Kansas, 2007); and Bruce G. Marcot and Jack Ward Thomas, *Of Spotted Owls, Old Growth, and New Politics: A History Since the Interagency Scientific Committee Report*, General Technical Report PNW-GTR-408 (U.S. Forest Service, Pacific Northwest Research Station, September 1997).

96. See Alagona, *After the Grizzly*; and Hennessy, *On the Backs of Tortoises*.

97. J. P. Linduska, "Endangered Species—The Federal Program," *The Elepaio: Journal of the Hawaii Audubon Society* 29, no. 1 (July 1968):1.

98. William J. Newman, Order 724, May 14, 1979, Nevada Division of Water Resources. Further restrictions on changing the point of diversion for existing water rights in order to protect the water level in Devils Hole were published in another order by the state engineer in 2008. See Tracy Taylor, Order 1197, November 4, 2008, Nevada Division of Water Resources; and Christopher J. Norment, *Relicts of a Beautiful Sea: Survival, Extinction, and Conservation in a Desert World* (Chapel Hill: University of North Carolina Press, 2014), 181–182.

99. William J. Newman, Ruling 2480, June 25, 1979, Nevada Division of Water Resources.

100. William J. Newman to John W. Burke, June 20, 1979, folder Amargosa drainage basin correspondence 1966, DWR-Carson City.

Chapter 6. After Victory

1. Devils Hole pupfish adult population count data, 1972–2015, Death Valley National Park, National Park Service, in author's possession; Devils Hole water data, 1937–2017, "USGS 362532116172700 230 S17 E50 36DC 1 Devils Hole (AM-4)," National Water Information System, U.S. Geological Survey, https://nwis.waterdata.usgs.gov/nwis/gwlevels/?site_no=362532116172700; and Point of Rocks Refuge and School Springs Refuge (Amargosa Pupfish Station) population data, 1980–1997, NDOW-Las Vegas.

2. "Pupfish Upheld," *New York Times*, June 19, 1976.

3. Deacon and Deacon Williams, "Ash Meadows and the Legacy of the Devils Hole Pupfish," 87.

4. For another analysis of the pupfish's legacy set in contrast to Deacon and Deacon Williams's early-1990s assessment, see Wilson, Hausner, and Brown, "Devils Hole Pupfish: Science in a Time of Crises," 387.

5. Devils Hole pupfish adult population count data, 1972–2015, Death Valley National Park, National Park Service, in author's possession.

6. Hoover Dam Refuge population data, 1999–2006, NDOW-Las Vegas.

7. Keith Rogers, "Count of Pupfish Shows Sharp Drop," *Las Vegas Review-Journal*, April 21, 2006.

8. On the genetic and ecological interpretations of the pupfish's declining population, see Andrew P. Martin et al., "Dramatic Shifts in the Gene Pool of a Managed Population of an Endangered Species May Be Exacerbated by High Genetic Load," *Conservation Genetics* 13 (2012):349–358; Wilson and Blinn, "Food Web Structure"; and Karly Feher, "Winter Dynamics in Mountain Lakes and Impacts of an Introduced Species to the Endangered Devils Hole Pupfish," master's thesis (University of Nevada, Reno, 2019).

9. This chapter title was inspired by an unusual source (for an environmental historian), a book by a political scientist examining international settlements and order after the conclusion of wars. See G. John Ikenberry, *After Victory: Institutions, Strategic Restraint, and the Rebuilding of Order after Major Wars* (Princeton, NJ: Princeton University Press, 2001).

10. Bill Toland, "In Desperate 1983, There Was Nowhere for Pittsburgh's Economy to Go but Up," *Pittsburgh Post Gazette*, December 23, 2012, https://www.post-gazette.com/business/businessnews/2012/12/23/In-desperate-1983-there-was-nowhere-for-Pittsburgh-s-economy-to-go-but-up/stories/201212230258.

11. Bruce Springsteen, "Youngstown," *The Ghost of Tom Joad*, Columbia, 1995. The literature on deindustrialization is large and crosses many fields. For my frames of reference, see Barry Bluestone and Bennett Harrison, *The Deindustrialization of America: Plant Closings, Commu-*

nity Abandonment, and the Dismantling of Basic Industry (New York: Basic Books, 1982); Judith Stein, *Pivotal Decade: How the United States Traded Factories for Finance in the Seventies* (New Haven, CT: Yale University Press, 2010); and Jefferson Cowie, *Stayin' Alive: The 1970s and the Last Days of the Working Class* (New York: New Press, 2010).

12. See, for example, Kahler Martinson to Director, BSFW, March 11, 1974, FWS-Sacramento; and Brown, *Recovering the Devils Hole Pupfish*, 293–294.

13. William R. Meyer to James Deacon, November 1, 1977, FWS-Las Vegas.

14. Karl Marx, "The Eighteenth Brumaire of Louis Napoleon," 1852, in *Karl Marx: Selected Writings*, ed. David McLellan (New York: Oxford University Press, 2000), 329.

15. Sid F. Cook and Cynthia D. Williams, *The Status and Future of Ash Meadows, Nye County, Nevada* (Carson City, NV: Office of the Attorney General, State of Nevada, May 19, 1982), V-2, V-3.

16. U.S. Department of Commerce, *1980 Census of Population*, vol. 1, PC80-1-A30 (Washington, D.C.: GPO, 1983), chap. A, part 30, p. 30-10.

17. On the growth of the Las Vegas metropolitan area, see Hal K. Rothman, *Neon Metropolis: How Las Vegas Started the Twenty-First Century* (New York: Routledge, 2002); and Mike Davis, "Las Vegas Versus Nature," in *Reopening the American West*, ed. Hal K. Rothman (Tucson: University of Arizona Press, 1998), 53–73. Preferred Equities Corporation's vision for Ash Meadows somewhat exemplifies historian Lincoln Bramwell's later concept of a "wilderburb." See Bramwell, *Wilderburbs: Communities on Nature's Edge* (Seattle: University of Washington Press, 2014).

18. Cook and Williams, *Status and Future of Ash Meadows*, V-8.

19. Donald Janson, "Land Hustlers in Las Vegas Thrive on Property-Hungry Vacationers," *New York Times*, September 14, 1972.

20. David Pauly, "Super-Salesman under Fire," *Newsweek*, February 18, 1980, 82; and Larry Kramer, "Las Vegas Developer Quietly Offers to Repay Small German Investors," *Washington Post*, March 25, 1978. On the history of Preferred Equities Corporation owner Leonard Rosen's previous company in Florida, see David E. Dodrill, *Selling the Dream: The Gulf American Corporation and the Building of Cape Coral, Florida* (Tuscaloosa: University of Alabama Press, 1993).

21. William Sweeney to Region 1 Director, FWS, March 23, 1982, FWS-Las Vegas; and "Emergency Determination of Endangered Status for Two Fish Species in Ash Meadows, Nevada," 47 Fed. Reg. 19995 (May 10, 1982).

22. On the history of land trusts in conservation, see Richard Brewer, *Conservancy: The Land Trust Movement in America* (Hanover, NH: Dartmouth College Press, 2003). Also see Sally K. Fairfax, Lauren Gwinn, Mary Ann King, Leigh Raymond, and Laura A. Watt, *Buying Nature: The Limits of Land Acquisition as a Conservation Strategy, 1780–2004* (Cambridge, MA: MIT Press, 2005).

23. Getting this deal completed is a saga worth many more words than I devote here. For fuller versions, see Brown, *Recovering the Devils Hole Pupfish*, 343–400; and Norment, *Relicts of a Beautiful Sea*, 151–191.

24. Department of the Interior, press release, "Ash Meadows to Become Newest National Wildlife Refuge," June 14, 1984, https://www.fws.gov/news/Historic/NewsReleases/1984/19840614.pdf.

25. Department of the Interior, press release, "Secretary Watt Outlines Conservation Policy to Avert Crises and to Service Present and Future Generations," March 23, 1981, box 266, Paul Laxalt Papers (83-01), Special Collections, University of Nevada, Reno.

26. Philip Shabecoff, "Deep Cuts Expected in Ecology Agencies," *New York Times*, March 12, 1981.

27. Congressional Research Service, *Land and Water Conservation Fund: Overview, Funding History, and Issues*, June 19, 2019, RL33531, p. 13.

28. Howell Raines, "States' Rights Move in West Influencing Reagan's Drive," *New York Times*, July 5, 1980.

29. Paul Laxalt to J. H. Hudgeons, September 9, 1983, box 605, Paul Laxalt Papers (83-01), Special Collections, University of Nevada, Reno.

30. Douglas B. Cornell Jr., notes from "DEVA field trip," [1984–85] box 7, Linda Greene Files, resource management collection, NPS-Cow Creek.

31. Death Valley National Monument, *Draft General Management Plan and Draft Environmental Impact Statement* (National Park Service, 1988), 39.

32. Sellars, *Preserving Nature in the National Parks*. See also R. Gerald Wright, *Wildlife Research and Management in the National Parks* (Urbana: University of Illinois Press, 1992); and Michael J. Yochim, *Protecting Yellowstone: Science and the Politics of National Park Management* (Albuquerque: University of New Mexico Press, 2013).

33. Tim Coonan to Ed Rothfuss, June 30, 1988, NPS-Pahrump.

34. James Deacon, reports and notes, 1974–1985, NPS-Pahrump.

35. James Deacon to Pete Sanchez, February 15, 1984, N1423, folder 2, box 29, central files, NPS-Cow Creek.

36. Pete Sanchez to Ed Rothfuss, March 15, 1984, N1423, folder 2, box 29, central files, NPS-Cow Creek.

37. James E. Deacon et al., *Devils Hole Pupfish Recovery Plan* (Portland, OR: U.S. Fish and Wildlife Service, 1980), 25–26.

38. The cost of Deacon's services found in a number of NPS documents in N1423, central files, NPS-Cow Creek. For inflation calculation, see U.S. Bureau of Labor Statistics, consumer price index inflation calculator tool, https://data.bls.gov/cgi-bin/cpicalc.pl ($4,000 in January 1985 is $9,807.70 in February 2020).

39. Rothman and Miller, *Death Valley National Park*, 136.

40. Sellers, *Preserving Nature in the National Parks*, 269–270.

41. "Superintendent's Annual Report: Death Valley National Monument," 1987, A2621, folder 25, central files, NPS-Cow Creek; and "Natural and Cultural Resources Programming Sheet," October 1983, N2623, box 8–13, central files, NPS-Cow Creek.

42. When an earthquake rocked Devils Hole in 1992 and lowered the pool level for a few months, the Park Service did contract for more regular pupfish population censuses for a twelve-month period in 1992–1993. See E. William Wischusen, "Final Report on Population Status of Devils Hole Pupfish (*Cyprinodon diabolis*) in Devils Hole Nevada," July 26, 1993, NPS-Cow Creek.

43. On the Las Vegas Valley Water District (after 1991, the Southern Nevada Water Authority), see Abraham Lustgarten, "The 'Water Witch': Pat Mulroy Preached Conservation while Backing Growth in Las Vegas," *ProPublica*, June 2, 2015, https://projects.propublica.org/killing-the-colorado/story/pat-mulroy-las-vegas-water-witch. Death Valley also found ways to trim its costs for water monitoring at Devils Hole during the 1980s. The monument had paid USGS to monitor Devils Hole's water level as well as the flow from Ash Meadows springs beginning in the 1970s, a cost that had risen to $20,000 by the mid-1980s. In 1989, Death Valley announced that the NPS Water Resources Division, based in Fort Collins, Colorado, would take over monitoring at Devils Hole. FWS would have to find a way to fund water monitoring on the Ash Meadows springs. As an agency, this change meant that NPS still picked up the tab at Devils Hole, but by eliminating the USGS contract the cost no longer came from Death Valley's base budget. See Brown, *Recovering the Devils Hole Pupfish*, 444–451.

44. James E. Deacon, "Report on Devils Hole," April 27, 1976, NPS-Pahrump.

45. Doug Threloff to "cooperators and interested parties," November 7, 1996, FWS-AMNWR.

46. James E. Deacon to Doug Threloff, December 23, 1996, NPS-Pahrump.

47. The format for this workshop had a precedent. Earlier in the decade, a separate series of "Devils Hole Workshops" began focused around research and monitoring of water resources in the Death Valley region and southern Nevada, generally.

48. NPS, Devils Hole Biological Workshop, background and agenda, April 3–4, 1998, NPS-Pahrump; and Michael S. Parker to Doug Threloff, April 10, 1998, NPS-Pahrump.

49. Blinn had worked extensively at Montezuma Well in Arizona. See, for example, Clay Runck and Dean W. Blinn, "Role of *Belostoma bakeri* (Heteroptera) in the Trophic Ecology of a Fishless Desert Spring," *Limnology and Oceanography* 39, no. 8 (1994):1800–1812.

50. The third collaborator on the bioenergetics study was David Herbst from University of California, Santa Barbara's Sierra Nevada Aquatic Research Laboratory.

51. For the major findings of the bioenergetics study, see Wilson and Blinn, "Food Web Structure." Interim and final reports by Blinn, Wilson, and Herbst to the Park Service are held at NPS-Pahrump.

52. Agenda and PowerPoint presentation from Devils Hole II Workshop, October 18, 2000, NDOW-Las Vegas.

53. Doug Threloff to Superintendent, Death Valley National Park (draft), December 27, 2000, NPS-Pahrump.

54. Doug Threloff, interview by Kevin C. Brown, March 6, 2014, Sacramento, California, transcript, author's possession, 12.

55. Endangered Species Act Amendments of 1978, Pub. L. 95-632, 92 Stat. 3751 at 3766 (1978). For a brief review and critique of the use of recovery teams, see William Burnham, Tom J. Cade, Alan Lieberman, J. Peter Jenny, and William R. Heinrich, "Hands-On Restoration," in *The Endangered Species Act at Thirty*, vol. 1, *Renewing the Conservation Promise*, eds. Dale D. Goble, J. Michael Scott, and Frank W. Davis (Washington, D.C.: Island Press, 2006), 237–246.

56. Steve Thompson to J. T. Reynolds, November 7, 2002, NPS-Pahrump.

57. Meeting summary, Devils Hole Recovery Team, July 9, 2003, NDOW-Las Vegas.

58. Assorted notes, Devils Hole Recovery Team 2002–2008, NDOW-Las Vegas; and notes and agendas, NPS-Pahrump.

59. William P. Van Liew to Cynthia Martinez, Terry Fisk, and others, April 14, 2004, NPS-Pahrump.

60. Deacon, et al., *Devils Hole Pupfish Recovery Plan*, 29–31.

61. Brown, *Recovering the Devils Hole Pupfish*, 417–418.

62. Brown, *Recovering the Devils Hole Pupfish*, 475.

63. Ash Meadows National Wildlife Refuge, "Amargosa Pupfish Station Management Plan for the School Springs and Point of Rocks Refugia," NPS-Pahrump.

64. Brown, *Recovering the Devils Hole Pupfish*, 477–479.

65. The invasive snail is the Red-rimmed melania (*Melanoides tuberculata*). See "*Melanoides tuberculata*," U.S. Geological Survey Nonindigenous Aquatic Species database, http://nas.er.usgs.gov/queries/FactSheet.aspx?speciesID=1037. On the origin and explosion of *Melanoides* in the Hoover Dam Refuge, see Brown, *Recovering the Devils Hole Pupfish*, 524–530.

66. The hybridization was first suspected when a researcher observed a pupfish with pelvic fins in the tank and was later confirmed with genetic testing. See Abraham Patrick Karam, "History and Development of Refuge Management for Devils Hole Pupfish (*Cyprinodon diabolis*) and an Ecological Comparison of Three Artificial Refuges" (master's thesis, Southern

Oregon University, 2005), 49–50, 67; and Andrew Martin, "Genetic Analysis of *Cyprinodon diabolis*: Hybridization with *C. nevandensis* in the Point of Rocks Refuge," unpublished report, 2005, NPS-Pahrump.

67. Susan M. Haig and Fred W. Allendorf, "Hybrids and Policy," in *The Endangered Species Act at Thirty*, vol. 2, *Conserving Biodiversity in Human-Dominated Landscapes*, eds. J. Michael Scott, Dale D. Goble, and Frank W. Davis (Washington, D.C.: Island Press, 2006), 150–163.

68. Point of Rocks refuge data logs, February and March 1996, FWS-AMNWR.

69. Henry Brean, "Death Valley Road Washes Out Again," *Las Vegas Review-Journal*, September 14, 2004.

70. Lindsey Treon Lyons, "Temporal and Spatial Variation in Larval Devils Hole Pupfish (*Cyprinodon diabolis*) Abundance and Associated Microhabitat Variables in Devils Hole, Nevada" (master's thesis, Southern Oregon University, 2005), 22–23.

71. Linda Greene to NPS, FWS, and NDOW personnel, September 23, 2004, NPS-Pahrump.

72. "Minutes of Multi-Agency Meeting Regarding September 2004 Flash Flood and Pupfish Take Incident at Devils Hole," October 8, 2004, NPS-Pahrump.

73. Launce Rake and Mary Manning, "Efforts to Study Endangered Fish Result in their Deaths," *Las Vegas Sun*, June 7, 2005; and Keith Rogers, "Flash Flood Proves Deadly to Rare Pupfish," *Las Vegas Review-Journal*, June 13, 2005.

74. On failure of captive propagation efforts in 2006–2007, see Brown, *Recovering the Devils Hole Pupfish*, 529–531.

75. Brown, *Recovering the Devils Hole Pupfish*, 532.

76. Linda Greene to J. T. Reynolds and David Ek, May 30, 2006, NPS-Pahrump; and Linda Greene to J. T. Reynolds, David Ek, and Mike R. Bower, August 9, 2006, NPS-Pahrump.

77. Mike R. Bower to Linda Greene, David Ek, and J. T. Reynolds, August 11, 2006, NPS-Pahrump.

78. Craig W. Thomas, *Bureaucratic Landscapes: Interagency Cooperation and the Preservation of Biodiversity* (Cambridge, MA: MIT Press, 2003), 1.

79. Richard A. Chase, FIRESCOPE*: A New Concept in Multiagency Fire Suppression Coordination*, General Technical Report PSW-40, Pacific Southwest Forest and Range Experiment Station, U.S. Forest Service (Berkeley, CA: Pacific Southwest Forest and Range Experiment Station, 1980), 1. On the Incident Command System, see also Dana Cole, "The Incident Command System: A 25-Year Evaluation by California Practitioners," National Fire Academy, February 2000, 209, https://www.hsdl.org/?view&did=454470; Jessica Jensen and Steven Thompson, "The Incident Command System: A Literature Review," *Disasters* 40, no. 1 (2016):158–182; Stephen J. Pyne, *California: A Fire Survey* (Tuscon: University of Arizona Press, 2016), esp. 34–37.

80. National Park Service, "Incident Command Structure for the Devils Hole Pupfish Recovery Effort," July 31, 2006, NPS-Pahrump.

81. In 2006, at the same time that Reynolds made this lasting contribution to pupfish management, he also was a vocal opponent of a plan by a George W. Bush—appointed deputy assistant secretary at the Interior Department to revise the types of recreation permitted in national parks to include the operation of snowmobiles and OHVS. Reynolds even found himself as a central character of a lengthy *Vanity Fair* article on the topic. See Michael Shnayerson, "Who's Ruining Our National Parks?" *Vanity Fair*, October 17, 2006, http://www.vanityfair.com/news/2006/06/nationalparks200606.

82. National Park Service, "Incident Command Structure for the Devils Hole Pupfish Recovery Effort," July 31, 2006, NPS-Pahrump. Other draft and final documents followed from this

proposal that formalized the Incident Command System at Devils Hole. See Brown, *Recovering the Devils Hole Pupfish*, 536–538.

83. Devils Hole Pupfish Incident Command Team, "2007 Annual Report on the Recovery and Management Actions for the Devils Hole Pupfish, *Cyprinodon diabolis*"; and "Devils Hole Pupfish Genetics Meeting, Fort Collins," August 28, 2007, James E. Deacon papers (hereafter Deacon papers).

84. John Wullschleger to Jon Jarvis, David Graber, and other NPS employees, January 3, 2007, NPS-Pahrump.

85. "Devils Hole Pupfish Weekly Activity Report for Week Ending January 12, 2007," NPS-Pahrump. See also, Michael R. Bower, "Pupfish Recovery Targets Food Availability at Devils Hole," *Park Science* 25, no. 1 (2008):26–27. On how the supplemental feeding program has changed since its implementation, see Brown, *Recovering the Devils Hole Pupfish*, 541–544.

86. A Death Valley National Park hydrology technician also works part time on Devils Hole related issues.

87. National Park Service, "Environmental Assessment: Devils Hole Long Term Ecosystem Monitoring Plan," Death Valley National Park, 2010; and National Park Service, "Devils Hole Ecosystem Monitoring Plan, Death Valley National Park," Natural Resources Report NPS/DEVA/NRR, 2011. Both documents held at NPS-Pahrump.

88. National Park Service, "Devils Hole Site Plan Environmental Assessment," Death Valley National Park, November 2009, NPS-Pahrump.

89. Funded projects, often led by graduate students as part of their thesis or dissertation requirements, are numerous and include Maria Christina Dzul et al., "A Simulation Model of the Devils Hole Pupfish Population Using Monthly Length-Frequency Distributions," *Population Ecology* 55 (2013):325–341; Mark B. Hausner et al., "Projecting the Effects of Climate Change and Water Management on Devils Hole Pupfish (*Cyprinodon diabolis*) Survival," *Ecohydrology* 9 (2016):560–573; and Feher, "Winter Dynamics in Mountain Lakes and Impacts of an Introduced Species to the Endangered Devils Hole Pupfish."

90. Ash Meadows National Wildlife Refuge, "Amargosa Pupfish Refugia Project," September 14, 2004, NDOW-Las Vegas; and Devils Hole Pupfish Incident Command Team, "2007 Annual Report on the Recovery and Management Actions for the Devils Hole Pupfish, *Cyprinodon diabolis*," Deacon papers.

91. Elizabeth Kolbert, *Under a White Sky: The Nature of the Future* (New York: Crown, 2021), 80.

92. Henry Brean, "New $4.5 Million Lab Battles for Life of Dwindling Devils Hole Pupfish," *Las Vegas Review-Journal*, December 1, 2013.

93. Southern Nevada Public Land Management Act of 1998, Pub. L. 105-263, 112 Stat. 2343 (1998); Jesse McKinley and Griffin Palmer, "Nevada Learns to Cash in on Sales of Federal Land," *New York Times*, December 3, 2007; and U.S. Bureau of Land Management, "Southern Nevada Public Land Management Act: Quick Facts and Program Statistics," https://www.blm.gov/sites/blm.gov/files/documents/files/SNPLMA_3%20PROGRAM%20STATISTICS.pdf, updated through July 31, 2019.

94. Lee H. Simons et al., "Study Proposal: Recovery and Husbandry of Devils Hole Pupfish Eggs from Devils Hole," July 15, 2013, FWS-AMNWR; and Olin G. Feuerbacher, Justin A. Mapula, and Scott A. Bonar, "Propagation of Hybrid Devils Hole Pupfish x Ash Meadows Amargosa Pupfish," *North American Journal of Aquaculture* 77 (2015):513–523.

95. Devils Hole Pupfish Incident Command Team, "Pupfish Egg Recovery from Devils Hole in 2013," June 2014, FWS-AMNWR.

96. Jason Bittel, "Brutal Beetles Kept World's Rarest Fish from Breeding—Until Now," *National Geographic*, March 1, 2019, https://www.nationalgeographic.com/animals/2019/03/endangered-devils-hole-pupfish-breeding-breakthrough/. On research at the Fish Facility, see also Joshua D. Sackett et al., "A Comparative Study of Prokaryotic Diversity and Physiochemical Characteristics of Devils Hole and the Ash Meadows Fish Conservation Facility, a Constructed Analog," *PloS ONE* 13, no. 3 (2018):e0194404.

97. Henry Brean, "Devils Hole Pupfish: Extinction Could Be Near with Just 35 Left," *Las Vegas Review-Journal*, April 25, 2013.

98. Death Valley National Park, press release, "Devils Hole Pupfish Population Reaches 136," April 3, 2019, https://www.nps.gov/deva/learn/news/devils-hole-pupfish-population-reaches-136.htm.

99. James E. Deacon, "Report on Devils Hole," April 27, 1976, NPS-Pahrump.

100. Sean Harris, "Nevada Department of Wildlife Devils Hole Pupfish Program 2010: Summary of Three-Year Project," 2010, 5, 17, NDOW-Las Vegas; and Devils Hole Pupfish Incident Command System, meeting notes, February 25, 2011, NPS-Pahrump.

Chapter 7. This Is Forever

1. Deacon et al., *Devils Hole Pupfish Recovery Plan*, 16–17. When the Devils Hole pupfish was incorporated into the Ash Meadows multispecies recovery plan ten years later, this language remained. See Sada, *Recovery Plan for the Endangered and Threatened Species of Ash Meadows, Nevada*, 36, 51.

2. Robert R. Miller to Carl L. Hubbs, June 22, 1937, folder 11, box 230, Hubbs papers, UCSD.

3. In the interest of full disclosure, I have collaborated elsewhere with Mark Hausner and the Park Service's Kevin Wilson on a short book chapter on the recent history and status of Devils Hole pupfish science and management. See Wilson, Hausner, and Brown, "The Devils Hole Pupfish: Science in a Time of Crises," in Propst et al., *Standing between Life and Extinction*, 379–389.

4. Mark B. Hausner, Kevin P. Wilson, D. Bailey Gaines, Francisco Suárez, G. Gary Scoppettone, and Scott W. Tyler, "Life in a Fishbowl: Prospects for the Endangered Devils Hole Pupfish (*Cyprinodon diabolis*) in a Changing Climate," *Water Resources Research* 50 (2014), doi.org/10.1002/2014WR015511. See also Mark B. Hausner, Kevin P. Wilson, D. Bailey Gaines, Francisco Suárez, and Scott W. Tyler, "The Shallow Thermal Regime of Devils Hole, Death Valley National Park," *Limnology and Oceanography: Fluids and Environments* 3 (2013):119–138, doi.org/10.1215/21573689-2372805; and Kayla D. Neuharth, "Modeling the Impacts of Sub-Seasonal Environmental Variability on the Devils Hole Pupfish" (master's thesis, University of Nevada, Reno, 2019).

5. Hausner et al., "Projecting the Effects of Climate Change."

6. Halford and Jackson, *Groundwater Characterization*, 74–75.

7. Ibid., 154–159. On the broader effects of groundwater pumping on biodiversity throughout southern and central Nevada, see James E. Deacon, Austin E. Williams, Cindy Deacon Willians, and Jack E. Williams, "Fueling Population Growth in Las Vegas: How Large-Scale Groundwater Withdrawal Could Burn Regional Biodiversity," *BioScience* 57, no. 8 (2007): 688–698.

8. Miller, "Correlation between Fish Distribution and Pleistocene Hydrography"; Hubbs and Miller, "The Zoological Evidence"; and Miller, "Speciation in Fishes."

9. Harry S. Truman, Presidential Proclamation 2961, January 17, 1952, in 17 Fed. Reg. 16 (January 23, 1952).

10. Robert R. Miller, "Speciation Rates in Some Fresh-Water Fishes of Western North America," in *Vertebrate Speciation: A University of Texas Symposium*, ed. W. Frank Blair (Austin: University of Texas Press, 1961), 548; and Robert R. Miller, "Coevolution of Deserts and Pupfishes (Genus *Cyprinodon*) in the American Southwest," in *Fishes in North American Deserts*, eds., Robert J. Naiman and David L. Soltz (New York: Wiley, 1981), 76.

11. Isaac J. Winograd et al., "Continuous 500,000-Year Climate Record from Vein Calcite in Devils Hole, Nevada," *Science* 258, no. 5080 (October 9, 1992):259. There is no evidence that the pupfish could have traveled to, let alone survived in, the cave before it collapsed and opened to the air.

12. Riggs and Deacon, "Connectivity in Desert Aquatic Ecosystems," 16.

13. Genetic methods are changing rapidly, but the recent papers explored here on the genetic evolutionary history of the pupfish are not the first in this line of investigation. For important earlier papers, see Anthony A. Echelle, "The Western North American Pupfish Clade (Cyprinodontidae: *Cyprinodon*): Mitochondrial DNA Divergence and Drainage History," in *Late Cenozoic Drainage History of the Southwestern Great Basin and Lower Colorado River Region: Geologic and Biotic Perspectives*, eds. M. C. Rheis, D. M. Hershler, and D. M. Miller, Geological Society of America, Special Paper 439 (2008):27–38; and Andrew P. Martin and Jennifer L. Wilcox, "Evolutionary History of Ash Meadows Pupfish (Genus *Cyprinodon*) Populations Inferred Using Microsatellite Markers," *Conservation Genetics* 5 (2004):769–782.

14. J. Michael Reed and Craig A. Stockwell, "Evaluating an Icon of Population Persistence: The Devils Hole Pupfish," *Proceedings of the Royal Society B: Biological Sciences* 281 (2014): 20141648, doi.org/10.1098/rspb.2014.1648; and Christopher H. Martin et al., "Diabolical Survival in Death Valley."

15. Christoper H. Martin et al., "Diabolical Survial in Death Valley," 8.

16. Sağlam, Baumsteiger, Smith, Lineres-Casenave, Nichols, O'Rourke, and Miller, "Phylogenetics Support an Ancient Common Origin of Two Scientific Icons."

17. On the back and forth between the Martin and Sağlam groups, see, İsmail K. Sağlam, Jason Baumsteiger, and Michael R. Miller, "Failure to Differentiate between Divergence of Species and Their Genes Can Result in Over-estimation of Mutation Rates in Recently Diverged Species," *Proceedings of the Royal Society B: Biological Sciences* 284 (2017):20170021, doi.org/10.1098/rspb.2017.0021; Christopher H. Martin, Sebastian Höhna, Jacob E. Crawford, Bruce J. Turner, Emilie J. Richards, and Lee H. Simons, "The Complex Effects of Demographic History on the Estimation of Substitution Rate: Concatenated Gene Analysis Results in No More Than Twofold Overestimation," *Proceedings of the Royal Society B: Biological Sciences* 284 (2017):20170537, doi.org/10.1098/rspb.2017.0537; Christopher H. Martin and Sebastian Höhna, "New Evidence for the Recent Divergence of Devils Hole Pupfish and the Plausibility of Elevated Mutation Rates in Endangered Taxa," *Molecular Ecology* 27 (2018):831-838, doi.org/10.1111/mec.14404; and İsmail K. Sağlam, Jason Baumsteiger, Matt J. Smith, Javier Lineres-Casenave, Andrew L. Nichols, Sean M. O'Rourke, and Michael R. Miller, "Best Available Science Still Supports an Ancient Common Origin of Devils Hole and Devils Hole Pupfish," *Molecular Ecology* 27 (2018):839–842, doi.org/10.1111.mec.14502.

18. Jonathan Amos, "Death Valley Fish a 'Recent Arrival,'" BBC *News*, January 27, 2016, https://www.bbc.com/news/science-environment-35420408.

19. Reed and Stockwell, "Evaluating an Icon," 3–4.

20. There are, however, reasons to be skeptical of the possibility that people introduced pupfish to Devils Hole, beyond the disagreements between Martin and Sağlam. No one as far as I can tell—not the scientists involved in this work or the Park Service—has ever seriously engaged either Western Shoshone or Southern Paiute groups whose ancestors lived in this region at the timing of the pupfish introduction. Nor have they sought to collaborate with historical anthropologists to see whether the idea of human introduction "makes sense" in the cultural context, not just in mutation rates. In an echo of the assumptions that led T. S. Palmer and the Death Valley Expedition not to ask native people about the names of the fish they collected for the 1891 survey, contemporary scientists have not examined their results in a way that invites perspectives beyond laboratory science.

21. Contemporaneous notes made during Martin's presentation, Desert Fishes Council, Furnace Creek, CA, November 16, 2018, in author's possession.

22. Hillary Rosner, "Attack of the Mutant Pupfish," *Wired*, November 19, 2012, https://www.wired.com/2012/11/mf-mutant-pupfish/. Christopher Norment also recounts the differing viewpoints of Andrew Martin and Kevin Wilson in *Relicts of a Beautiful Sea*, 142–145.

23. Pittman, *Cat Tale*.

24. Commenting on the two initial papers by the Martin and Sağlam groups, authors of a review of conservation genetics in desert fishes noted that both researchers "expressed high levels of confidence in their (mutually exclusive) results." They added, "Because critical decisions about whether and how to implement assisted gene flow or hatchery programs depend on the evolutionary provenance of Devils Hole pupfish, independent scrutiny of the original data, including critical evaluation of how missing data affected the conclusions, must be conducted prior to the development and implementation of a genetic management plan." See Thomas F. Turner et al., "Conservation Genetics of Desert Fishes in the Genomics Age," in *Standing Between Life and Extinction*, eds. Propst et al., 210.

25. Norment, *Relicts of a Beautiful Sea*, 147.

26. "Do We Have Unlimited Resources to Save the Devils Hole Pupfish?" *Las Vegas Review-Journal*, April 22, 2018, https://www.reviewjournal.com/opinion/editorials/editorial-do-we-have-unlimited-resources-to-save-the-devils-hole-pupfish/.

27. On expenditures by FWS and the National Marine Fisheries Service on endangered species recovery, see Noah Greenwald, Brett Hartl, Loyal Mehrhoff, and Jamie Pang, "Shortchanged: Funding Needed to Save America's Most Endangered Species," Center for Biological Diversity, December 2016, 10–11, https://www.biologicaldiversity.org/programs/biodiversity/pdfs/Shortchanged.pdf. These totals do not include funding from other agencies, such as NPS, but do provide some scale of different recovery efforts. The Devils Hole pupfish, in 2014, did not make the top thirty-five.

28. J. P. Linduska, handwritten note, n.d., stapled to front of James T. McBroom to Director, BSFW, August 24, 1970, FWS-Las Vegas.

29. Wes Jackson, "The Necessity and Possibility of an Agriculture Where Nature is the Measure," *Conservation Biology* 22, no. 6 (2008):1376–1377. For a further discussion of Jackson's idea, see Wendell Berry, *Our Only World: Ten Essays* (Berkeley, CA: Counterpoint, 2015), 115.

Bibliography

Archival Records

Deacon, James E., personal papers

Ira La Rivers collection, California Academy of Sciences, Library and Archives, San Francisco, California

National Archives and Records Administration

College Park, Maryland (NARA-College Park)

Records of the U.S. Fish and Wildlife Service (RG 22)

Records of the National Park Service (RG 79)

San Bruno, California (NARA-San Bruno)

Records of the Bureau of Land Management (RG 49)

Copies of records ordered via email or viewed online (NARA-Online)

Center for Legislative Archives, Records of the U.S. Senate (RG 46)

General Land Office, Land Entry Case Files (RG 49)

Records of the Department of State (RG 59)

Watergate Special Prosecution Force, Campaign Contribution Task Force (RG 460)

National Park Service

Death Valley National Park Archives, Cow Creek, California (NPS-Cow Creek)

Death Valley National Park office, Pahrump, Nevada (NPS-Pahrump)

Nevada Department of Wildlife

Las Vegas, Nevada office (NDOW-Las Vegas)

Nevada Division of Water Resources

Carson City, Nevada office (DWR-Carson City)

Online records, http:/water.nv.gov

Theodore Sherman Palmer papers, Huntington Library, San Marino, California

Pister, Edwin P. (Phil), personal papers

Smithsonian Institution, Fishes Division, National Museum of Natural History

University of California, Berkeley, Museum of Vertebrate Zoology

Carol J. James field notes collection (MVZA.MSS.0458)

Clinton Hart Merriam papers (MVZA.MSS.0281)

University of California, San Diego, Special Collections and Archives, Geisel Library

Scripps Institution of Oceanography Archives

Carl Leavitt Hubbs papers (81–18)

University of Michigan, Fishes Division, Museum of Zoology

Field notes collection

Robert Rush Miller papers

Specimen accession records

University of Nevada, Las Vegas, Special Collections and Archives, Lied Library
Desert Pupfish collection (MS 20)
University of Nevada, Reno, Special Collections and University Archives, Mathewson-IGT Knowledge Center
Alan Bible papers (NC01)
Foresta Institute for Ocean and Mountain Studies records (2001-01)
Paul Laxalt papers (83-01)
James D. Yoakum papers (2013-27)
U.S. Bureau of Land Management
Reno, Nevada office (BLM-Reno)
U.S. Fish and Wildlife Service
Ash Meadows National Wildlife Refuge office (FWS-AMNWR)
Las Vegas, Nevada office (FWS-Las Vegas)
Sacramento, California office (FWS-Sacramento)

Oral History Interviews

Conducted by author. Digital recordings deposited (with transcripts) at Death Valley National Park, Cow Creek archives:
James E. Deacon, Henderson, Nevada, April 3, 2014
Douglas Threloff, Sacramento, California, March 6, 2014

Pister, Edwin Philip. "Preserving Native Fishes and Their Ecosystems: A Pilgrim's Progress, 1950s–Present," an oral history conducted by Ann Lage in 2007–2008, Regional Oral History Office, The Bancroft Library, University of California, Berkeley, 2009.

Published Records

References to newspapers, court cases, legislation, death certificates, news releases, online databases, and the Federal Register appear only in the endnotes, not in this bibliography.

Abbey, Edward. *The Journey Home: Some Words in Defense of the American West.* New York: Dutton, 1977.
Abdoun, Hany. "Ira John La Rivers (1915–1977)." California Academy of Sciences. http://researcharchive.calacademy.org/research/library/special/bios/LaRivers.pdf.
Alagona, Peter S. *After the Grizzly: Endangered Species and the Politics of Place in California.* Berkeley: University of California Press, 2013.
——. "Species Complex: Classification and Conservation in American Environmental History." *Isis* 107, no. 4 (2016):738–761.
Andrés, Benny J., Jr. *Power and Control in the Imperial Valley: Nature, Agribusiness, and Workers on the California Borderland, 1900–1940.* College Station: Texas A&M University Press, 2015.
Andrews, Richard N. L. *Managing the Environment, Managing Ourselves: A History of American Environmental Policy.* 2nd ed. New Haven, CT: Yale University Press, 2006.
Asma, Stephen T. *Stuffed Animals and Pickled Heads: The Culture and Evolution of Natural History Museums.* New York: Oxford University Press, 2001.
Bailey, Vernon O. "Into Death Valley 50 Years Ago." *Westways* (December 1940):8–11.
Barlett, Donald L., and James B. Steele. *Empire: The Life, Legend, and Madness of Howard Hughes.* New York: Norton, 1979. Reprinted as *Howard Hughes: His Life and Madness.* New York: Norton, 2004.

Barnosky, Anthony D., Nicholas Matzke, Susumu Tomiya, Guinevere O. U. Wogan, Brian Swartz, Tiago B. Quental, Charles Marshall, et al. "Has the Earth's Sixth Mass Extinction Already Arrived?" *Nature* 471 (March 3, 2011):51–57.

Barrow, Mark V., Jr. *Nature's Ghosts: Confronting Extinction from the Age of Jefferson to the Age of Ecology*. Chicago: University of Chicago Press, 2009.

Baugh, Thomas M., and James E. Deacon. "Evaluation of the Role of Refugia in Conservation Efforts for the Devils Hole Pupfish, *Cyprinodon diabolis* Wales." *Zoo Biology* 7 (1988): 351–58.

Baumberger, J. Percy. "A History of Biology at Stanford University." *Bios* 25, no. 3 (1954):123–147.

Bean, Michael J. "Biological Diversity and the Refuge System: Beyond the Endangered Species Act in Fish and Wildlife Management." In *Transactions of the Fifty-Seventh North American Wildlife and Natural Resources Conference*, edited by Richard E. McCabe, 577–580. Washington, D.C.: Wildlife Management Institute, 1992.

Bean, Tarleton H. *Directions for Collecting and Preserving Fish*. Washington, D.C.: GPO, 1881.

Beissinger, Steven R. "Digging the Pupfish Out of Its Hole: Risk Analyses to Guide Harvest of Devils Hole Pupfish for Captive Breeding." *PeerJ Life and Environment* 2, no. 1 (2014):e549. doi.org/10.7717/peerj.549.

Berry, Wendell. *Our Only World: Ten Essays*. Berkeley: Counterpoint, 2015.

Bittel, Jason. "Brutal Beetles Kept World's Rarest Fish from Breeding—Until Now." *National Geographic*, March 1, 2019. https://www.nationalgeographic.com/animals/2019/03/endangered-devils-hole-pupfish-breeding-breakthrough/.

Black, Megan. *The Global Interior: Mineral Frontiers and American Power*. Cambridge, MA: Harvard University Press, 2018.

Blankenbuehler, Paige. "How a Tiny Endangered Species Put a Man in Prison." *High Country News*, April 15, 2019. https://www.hcn.org/issues/51.6/endangered-species-how-a-tiny-endangered-species-put-a-man-in-prison.

Bluestone, Barry, and Bennett Harrison. *The Deindustrialization of America: Plant Closings, Community Abandonment, and the Dismantling of Basic Industry*. New York: Basic Books, 1982.

Blunt, Wilfrid. *Linnaeus: The Compleat Naturalist*. Revised edition. Princeton, NJ: Princeton University Press, 2002.

Bower, Michael R. "Pupfish Recovery Targets Food Availability at Devils Hole." *Park Science* 25, no. 1 (2008):26–27.

Bowker, Geoffrey C., and Susan Lee Star. *Sorting Things Out: Classification and Its Consequences*. Cambridge, MA: MIT Press, 1999.

Bowler, Peter J. *Evolution: A History of the Idea*. 25th anniversary edition. Berkeley: University of California Press, 2009.

Bramwell, Lincoln. *Wilderburbs: Communities on Nature's Edge*. Seattle: University of Washington Press, 2014.

Brewer, Richard. *Conservancy: The Land Trust Movement in America*. Hanover, NH: Dartmouth College Press, 2003.

Brown, Kate. *Dispatches from Dystopia: Histories of Places Not Yet Forgotten*. Chicago: University of Chicago Press, 2015.

Brown, Kevin C. "'An Exceedingly Simple, Little Ecosystem': Devils Hole, Endangered Species Conservation, and Scientific Environments." *Notes and Records: The Royal Society Journal of the History of Science* 75, no. 2 (2021):239–257.

———. "The 'National Playground Service' and the Devils Hole Pupfish." *Forest History Today* (Spring 2017):35–40.

———. *Recovering the Devils Hole Pupfish: An Environmental History.* National Park Service: Death Valley National Park, 2017.

Brues, Charles T. "Studies of the Fauna of Hot Springs in the Western United States and the Biology of Thermophilous Animals." *Proceedings of the American Academy of Arts and Sciences* 63, no. 4 (1928):139–228.

Bugliosi, Vincent, with Curt Gentry. *Helter Skelter: The True Story of the Manson Murders.* New York: Norton, 1974.

Bulkley, Robert J., Jr. *At Close Quarters: PT Boats in the Pacific.* Washington, D.C.: GPO, 1962.

Burkhead, Noel M. "Extinction Rates in North American Freshwater Fishes, 1900–2010." *BioScience* 62, no. 9 (2012):798–808.

Burnham, Philip. *Indian Country, God's Country: Native Americans and the National Parks.* Washington, D.C.: Island Press, 2000.

Burnham, William, Tom J. Cade, Alan Lieberman, J. Peter Jenny, and William R. Heinrich. "Hands-on Restoration." In *The Endangered Species Act at Thirty.* Vol. 1, *Renewing the Conservation Promise,* edited by Dale D. Goble, J. Michael Scott, and Frank W. Davis, 237–246. Washington, D.C.: Island Press, 2006.

Cahalan, James M. *Edward Abbey: A Life.* Tucson: University of Arizona Press, 2001.

Cameron, Jenks. *The Bureau of Biological Survey: Its History, Activities, and Organization.* Institute for Government Research, Service Monographs of the United States Government 54. Baltimore: Johns Hopkins Press, 1929.

Cashner, Robert Charles, Gerald R. Smith, Gifford Hubbs Miller, Frances Miller Cashner, and Barry Chernoff. "Robert Rush Miller (1916–2003)." *Copeia* 2011, no. 2 (June 28, 2011):342–347.

Cawley, R. McGreggor. *Federal Land, Western Anger: The Sagebrush Rebellion and Environmental Politics.* Lawrence: University Press of Kansas, 1993.

Ceballos, Gerardo, Paul R. Ehrlich, and Rodolfo Dirzo. "Biological Annihilation via the Ongoing Sixth Mass Extinction Signaled by Vertebrate Population Losses and Declines." *PNAS* 114, no. 30 (July 2017):E6089-E6096. doi.org/10.1073/pnas.1704949114.

Chase, Richard A. *FIRESCOPE: A New Concept in Multiagency Fire Suppression Coordination.* General Technical Report PSW-40. Pacific Southwest Forest and Range Experiment Station, U.S. Forest Service. Berkeley, CA: Pacific Southwest Forest and Range Experiment Station, 1980.

Clark, Eugenie. *Lady with a Spear.* New York: Harper, 1951.

Cole, Dana. "The Incident Command System: A 25-Year Evaluation by California Practitioners." National Fire Academy, February 2000. https://www.hsdl.org/?view&did=454470.

Committee on Rare and Endangered Wildlife Species. *Rare and Endangered Fish and Wildlife of the United States.* Resource Publication 34. Washington, D.C.: Bureau of Sport Fisheries and Wildlife, July 1966.

Congressional Research Service. *National Monuments and the Antiquities Act.* CRS Report R41330, November 30, 2018.

———. *Land and Water Conservation Fund: Overview, Funding History, and Issues.* CRS Report RL33531, June 19, 2019.

Cook, Sid F., and Cynthia D. Williams. *The Status and Future of Ash Meadows, Nye County, Nevada.* Carson City, NV: Office of the Attorney General, State of Nevada, May 19, 1982.

Coues, Elliot. *Key to North American Birds.* Boston: Estes and Lauriat, 1884.

———. Letter to the editor, "Trinomials Are Necessary," *The Auk* 1, no. 2 (April 1884):197–198.

Coville, Frederick Vernon. "The Panamint Indians of California." *American Anthropologist* 5, no. 4 (October 1892):351–362.

Cowie, Jefferson. *Stayin' Alive: The 1970s and the Last Days of the Working Class*. New York: New Press, 2010.

Curtin, Charles G. "The Evolution of the U.S. National Wildlife Refuge System and the Doctrine of Compatibility." *Conservation Biology* 7, no. 1 (March 1993):29–38.

Davis, Mike. "Las Vegas Versus Nature." In *Reopening the American West*, edited by Hal K. Rothman, 53–73. Tucson: University of Arizona Press, 1998.

Davison, Robert P., Alessandra Falucci, Luigi Maiorano, and J. Michael Scott. "The National Wildlife Refuge System." In *The Endangered Species Act at Thirty*. Vol. 1, *Renewing the Conservation Promise*, edited by Dale D. Goble, J. Michael Scott, and Frank W. Davis, 90–100. Washington, D.C.: Island Press, 2005.

Dayan, Tamar, and Bella Galil. "Natural History Collections as Dynamic Research Archives." In *Stepping in the Same River Twice: Replication in Biological Research*, edited by Ayelet Shavit and Aaron M. Ellison, 55–79. New Haven, CT: Yale University Press, 2017.

Deacon, James E., and Cynthia Deacon Williams. "Ash Meadows and the Legacy of the Devils Hole Pupfish." In *Battle Against Extinction: Native Fish Management in the American West*, edited by W. L. Minckley and James E. Deacon, 69–87. Tucson: University of Arizona Press, 1991.

Deacon, James E., Dale Lockard, Osborne Casey, Gail Kobetich, Jon Radtke, Herb Gunther, and Dave Soltz. *Devils Hole Pupfish Recovery Plan*. Portland, OR: U.S. Fish and Wildlife Service, 1980.

Deacon, James E., Austin E. Williams, Cindy Deacon Williams, and Jack E. Williams. "Fueling Population Growth in Las Vegas: How Large-Scale Groundwater Withdrawal Could Burn Regional Biodiversity." *BioScience* 57, no. 8 (September 2007):688–698.

Deacon Williams, Cindy. "James E. Deacon." In *Standing between Life and Extinction: Ethics and Ecology of Conserving Aquatic Species in North American Deserts*, edited by David L. Propst, Jack E. Williams, Kevin R. Bestgen, and Christopher W. Hoagstrom, 22-26. Chicago: University of Chicago Press, 2020.

Death Valley National Monument. *Draft General Management Plan and Draft Environmental Impact Statement*. National Park Service, 1988.

Delbourgo, James. *Collecting the World: Hans Sloane and the Origins of the British Museum*. Cambridge, MA: Harvard University Press, 2017.

Deur, Douglas, and Deborah Confer. *People of Snowy Mountain, People of the River: A Multi-Agency Ethnographic Overview and Compendium Relating to Tribes Associated with Clark County, Nevada*. National Park Service, Pacific West Region, Social Science Series 2012-01. 2012.

Dieter, Cheryl A., Molly A. Maupin, Rodney R. Caldwell, Melissa A. Harris, Tamara I. Ivahnenko, John K. Lovelace, Nancy L. Barber, and Kristin S. Linsey. *Estimated Use of Water in the United States in 2015*. Circular 1441. Reston, VA: U.S. Geological Survey, 2018.

Dilsaver, Lary M. "Not of National Significance: Failed National Park Proposals in California." *California History* 85, no. 2 (2008):4–23.

———, ed. *America's National Park System: The Critical Documents*. 2nd edition. New York: Rowman and Littlefield, 2016.

Dodrill, David E. *Selling the Dream: The Gulf American Corporation and the Building of Cape Coral, Florida*. Tuscaloosa: University of Alabama Press, 1993.

Dudley, W.W., Jr., and J.D. Larson. *Effect of Irrigation Pumping on Desert Pupfish Habitats in Ash Meadows, Nye County, Nevada.* Geological Survey Professional Paper 927. Washington, D.C.: GPO, 1976.

Dunlap, Thomas R. *Saving America's Wildlife: Ecology and the American Mind, 1850–1990.* Princeton, NJ: Princeton University Press, 1988.

Dunn, J. Richard. "Charles Henry Gilbert (1859–1928): Pioneer Ichthyologist of the American West." In *Collection Building in Ichthyology and Herpetology*, edited by Theodore W. Pietsch and William D. Anderson Jr., 265–278. American Society of Ichthyologists and Herpetologists Special Publication 3. Lawrence, KS: Allen Press, 1997.

Dzul, Maria Christina, Stephen James Dinsmore, Michael Carl Quist, Daniel Bailey Gaines, Kevin Patrick Wilson, Michael Roy Bower, and Philip Michael Dixon. "A Simulation Model of the Devils Hole Pupfish Population Using Monthly Length-Frequency Distributions." *Population Ecology* 55 (2013):325–341.

Echelle, Anthony A. "The Western North American Pupfish Clade (Cyprinodontidae: *Cyprinodon*): Mitochondrial DNA Divergence and Drainage History." In *Late Cenozoic Drainage History of the Southwestern Great Basin and Lower Colorado River Region: Geologic and Biotic Perspectives*, edited by M.C. Rheis, D.M. Hershler, and D.M. Miller, 27–38. Geological Society of America, Special Paper 439. 2008.

"Editorial News and Notes." *Copeia* 1946, no. 3 (October 20, 1946):180.

Edwards, Robert J. "Carl Leavitt Hubbs and Robert Rush Miller." In *Standing between Life and Extinction: Ethics and Ecology of Conserving Aquatic Species in North American Deserts*, edited by David L. Propst, Jack E. Williams, Kevin R. Bestgen, and Christopher W. Hoagstrom, 13–16. Chicago: University of Chicago Press, 2020.

Elliott, Gary E. *Senator Alan Bible and the Politics of the New West.* Reno: University of Nevada Press, 1994.

Energy Research and Development Administration. *Nevada Test Site, Nye County, Nevada: Final Environmental Impact Statement.* September 1977.

Fairfax, Sally K., Lauren Gwin, Mary Ann King, Leigh Raymond, and Laura A. Watt. *Buying Nature: The Limits of Land Acquisition as a Conservation Strategy, 1780–2004.* Cambridge, MA: MIT Press, 2005.

Farnham, Timothy J. *Saving Nature's Legacy: Origins of the Idea of Biological Diversity.* New Haven, CT: Yale University Press, 2007.

Feher, Karly. "Winter Dynamics in Mountain Lakes and Impacts of an Introduced Species to the Endangered Devils Hole Pupfish." Master's thesis, University of Nevada, Reno, 2019.

Feuerbacher, Olin G., Justin A. Mapula, and Scott A. Bonar. "Propagation of Hybrid Devils Hole Pupfish x Ash Meadows Amargosa Pupfish." *North American Journal of Aquaculture* 77 (2015):513–523.

Fiero, G. William, and G.B. Maxey. *Hydrogeology of the Devils Hole Area, Ash Meadows, Nevada.* Reno: Center for Water Resources Research, Desert Research Institute, University of Nevada System, June 1970. https://archive.org/details/hydrogeologyofdeoofier/mode/2up.

Fine, Lisa M. *The Story of Reo Joe: Work, Kin, and Community in Autotown, USA.* Philadelphia: Temple University Press, 2004.

Foucault, Michel. *The Order of Things: An Archaeology of the Human Sciences.* New York: Vintage, 1973.

Fowler, Catherine S. "What's in a Name? Some Southern Paiute Names for Mojave Desert Springs as Keys to Environmental Perception." In *Conference Proceedings, Spring-Fed Wet-*

lands: Important Scientific and Cultural Resources of the Intermountain Region, May 7–9, 2002, edited by D. W. Sada and S. E. Sharpe. DRI Report 41201. Reno: Desert Research Institute, 2004.

Gilbert, Charles Henry. "Report on Fishes..." In *The Death Valley Expedition: A Biological Survey of Parts of California, Nevada, Arizona, and Utah*, part 2, 229–234. North American Fauna 7. Washington, D.C.: Division of Ornithology and Mammalogy, U.S. Department of Agriculture, GPO, 1893.

Gilbert, Charles Henry, and Norman Bishop Scofield. "Notes on the Fishes from the Colorado Basin in Arizona." *Proceedings of the United States National Museum* 20, 487–499. Washington, D.C.: GPO, 1898.

Girard, Charles. "Ichthyology of the Boundary." In *United States and Mexican Boundary Survey, under the Order of Lieut. Col. W. H. Emory, Major First Cavalry, and United States Commissioner*. Vol. 2, part 6, p. 68. Washington, D.C.: Cornelius Wendell, 1859. http://hdl.handle.net/2027/miun.afk4546.0002.006.

Glennon, Robert. *Water Follies: Groundwater Pumping and the Fate of America's Fresh Waters*. Washington, D.C.: Island Press, 2002.

Goetzmann, William H. *Exploration and Empire: The Explorer and the Scientist in the Winning of the American West*. New York: Knopf, 1966.

Green, Donald E. *Land of the Underground Rain: Irrigation on the High Texas Plains, 1910–1970*. Austin: University of Texas Press, 1973.

Green, Michael S. *Nevada: A History of the Silver State*. Reno: University of Nevada Press, 2015.

Greenwald, Noah, Brett Hartl, Loyal Mehrhoff, and Jamie Pang. "Shortchanged: Funding Needed to Save America's Most Endangered Species." Tucson, AZ: Center for Biological Diversity, December 2016. https://www.biologicaldiversity.org/programs/biodiversity/pdfs/Shortchanged.pdf.

Grinnell, Hilda Wood. "Joseph Grinnell: 1877–1939." *The Condor* 42, no. 1 (1940):3–34.

Grinnell, Joseph. "The Museum Conscience." *Museum Work* 4 (September–October 1921):62–63.

Grinnell, Joseph, and Tracy I. Storer. "Animal Life as an Asset of National Parks." *Science* 44, no. 1133 (September 15, 1916):375–380.

Haig, Susan M., and Fred W. Allendorf. "Hybrids and Policy." In *The Endangered Species Act at Thirty*. Vol. 2, *Conserving Biodiversity in Human-Dominated Landscapes*, edited by J. Michael Scott, Dale D. Goble, and Frank W. Davis, 150–163. Washington, D.C.: Island Press, 2006.

Halford, Keith J., and Tracie R. Jackson. *Groundwater Characterization and Effects of Pumping in the Death Valley Regional Groundwater Flow System, Nevada and California, with Special Reference to Devils Hole*. Professional Paper 1863. Reston, VA: U.S. Geological Survey, 2020.

Halverson, Anders. *An Entirely Synthetic Fish: How Rainbow Trout Beguiled America and Overran the World*. New Haven, CT: Yale University Press, 2010.

Hanemann, W. Michael. "The Central Arizona Project." Working Paper 937. Department of Agricultural and Resource Economics and Policy, Division of Agricultural and Natural Resources, University of California, Berkeley, October 2002. https://escholarship.org/uc/item/87k436cf.

Harmon, David. "The Antiquities Act and Nature Conservation." In *The Antiquities Act: A Century of American Archeology, Historic Preservation, and Nature Conservation*, edited by David Harmon, Francis P. McManamon, and Dwight T. Pitcaithley, 199–215. Tucson: University of Arizona Press, 2006.

Harrill, James R. *Pumping and Ground-Water Storage Depletion in Las Vegas Valley, Nevada, 1955-74*. Water-Resources Bulletin 44, pp. 37–43. State of Nevada Department of Conservation and Natural Resources, Division of Water Resources, 1976.

Harrison, Sylvia. "The Historical Development of Nevada Water Law." *University of Denver Water Law Review* 5, no. 1 (2001):148–182.

Harvey, Mark W. T. *A Symbol of Wilderness: Echo Park and the American Conservation Movement*. Albuquerque: University of New Mexico Press, 1994.

Hausner, Mark B., Kevin P. Wilson, D. Bailey Gaines, Francisco Suárez, G. Gary Scoppettone, and Scott W. Tyler. "Life in a Fishbowl: Prospects for the Endangered Devils Hole Pupfish (*Cyprinodon diabolis*) in a Changing Climate." *Water Resources Research* 50 (2014):1–15. doi .org/10.1002/2014WR015511.

———. "Projecting the Effects of Climate Change and Water Management on Devils Hole Pupfish (*Cyprinodon diabolis*) Survival." *Ecohydrology* 9 (2016):560–573.

Hausner, Mark B., Kevin P. Wilson, D. Bailey Gaines, Francisco Suárez, and Scott W. Tyler. "The Shallow Thermal Regime of Devils Hole, Death Valley National Park." *Limnology and Oceanography: Fluids and Environments* 3 (2013):119–138.

Hays, Samuel P. *Conservation and the Gospel of Efficiency: The Progressive Conservation Movement, 1890–1920*. Cambridge, MA: Harvard University Press, 1959.

Heise, Ursula K. *Imagining Extinction: The Cultural Meaning of Endangered Species*. Chicago: University of Chicago Press, 2016.

Hennessy, Elizabeth. *On the Backs of Tortoises: Darwin, the Galápagos, and the Fate of an Evolutionary Eden*. New Haven: Yale University Press, 2019.

———. "Saving Species: The Co-Evolution of Tortoise Taxonomy and Conservation in the Galápagos Islands." *Environmental History* 25 (2020):263–286.

Hornaday, William T. *Our Vanishing Wildlife: Its Extermination and Preservation*. 1913. Reprint, New York: Arno and New York Times, 1970.

Hubbs, Carl L. "Criteria for Subspecies, Species and Genera, as Determined by Researches on Fishes," *Annals of the New York Academy of Sciences* 44, no. 2 (1943):109-121.

———. "Fishes from the Big Bend Region of Texas." *Transactions of the Texas Academy of Science* 23 (1940):3–12.

———. "History of Ichthyology in the United States after 1850." *Copeia* 1964, no. 1 (March 26, 1964):42–60.

———. *Studies of the Fishes of the Order Cyprinodontes VI*. University of Michigan Museum of Zoology Miscellaneous Publications 16. Ann Arbor: University of Michigan, July 9, 1926.

Hubbs, Carl L., and Karl F. Lagler. "Fishes of Isle Royale, Lake Superior, Michigan." *Papers of the Academy of Science, Arts, and Letters* 33, 1947, pp. 73–133. Ann Arbor: University of Michigan Press, 1949.

Hubbs, Carl L., and Robert R. Miller. "The Zoological Evidence: Correlation between Fish Distribution and Hydrographic History in the Desert Basins of the Western United States." In *The Great Basin, with Emphasis on Glacial and Postglacial Times*. Bulletin of the University of Utah 38, no. 20 (1948):17–166.

Hubbs, Carl L., and J. Clark Salyer. "Fish Conditions in Iowa Lakes, with Recommendations for the Fish Management of These Lakes." Report 209. Ann Arbor: Institute for Fisheries Research, Michigan Department of Conservation and University of Michigan, May 2, 1933.

Hubbs, Carl L., and Orthello L. Wallis. "The Native Fish Fauna of Yosemite National Park and Its Preservation." *Yosemite Nature Notes* 27, no. 12 (December 1948):131–144.

Humphreys, A. A. *Preliminary Report Concerning Explorations and Surveys Principally in Nevada and Arizona... Conducted under the Immediate Direction of 1st Lieut. George M. Wheeler, Corps of Engineers, 1871.* Washington, D.C.: GPO, 1872.

Hundley, Norris, Jr. *The Great Thirst: Californians and Water: A History.* Revised edition. Berkeley: University of California Press, 2001.

———. "The 'Winters' Decision and Indian Water Rights: A Mystery Reexamined." *Western Historical Quarterly* 13, no. 1 (1982):17–42.

Ikenberry, G. John. *After Victory: Institutions, Strategic Restraint, and the Rebuilding of Order after Major Wars.* Princeton, NJ: Princeton University Press, 2001.

Ingram, B. Lynn, and Frances Malamud-Roam. *The West without Water: What Past Floods, Droughts, and Other Climatic Clues Tell Us about Tomorrow.* Berkeley: University of California Press, 2013.

Intergovernmental Science-Policy Platform on Biodiversity and Ecosystem Services, IPBES. *Summary for Policymakers of the Global Assessment Report on Biodiversity and Ecosystem Services.* 2019. https://zenodo.org/record/3553579.

Isenberg, Andrew C. *The Destruction of the Bison: An Environmental History, 1750–1920.* Cambridge: Cambridge University Press, 2000.

Jackson, Wes. "The Necessity and Possibility of an Agriculture Where Nature Is the Measure." *Conservation Biology* 22, no. 6 (2008):1376–1377.

Jacoby, Karl. *Crimes Against Nature: Squatters, Poachers, Thieves, and the Hidden History of American Conservation.* Berkeley: University of California Press, 2001.

James, Carol Jeanne. "Aspects of the Ecology of the Devils Hole Pupfish, *Cyprinodon diabolis* Wales." Master's thesis, University of Nevada, Las Vegas, 1969.

Jensen, Jessica, and Steven Thompson. "The Incident Command System: A Literature Review." *Disasters* 40, no. 1 (2016):158–182.

Johnson, Leroy, and Jean Johnson, eds. *Escape from Death Valley: As Told by William Lewis Manly and Other '49ers.* Reno: University of Nevada Press, 1987.

Jordan, David Starr. "Charles Henry Gilbert." *Science* 67, no. 1748 (June 29, 1928):644–645.

———. *A Guide to the Study of Fishes.* Vol. 1. New York: Henry Holt, 1905.

———. "The Origin of Species through Isolation," *Science* 22, no. 566 (November 3, 1905):545–562.

Jordan, David Starr, and Barton Warren Evermann. *The Fishes of North and Middle America: A Descriptive Catalogue of the Species of Fish-Like Vertebrates Found in the Waters of North America, North of the Isthmus of Panama,* part 1. Bulletin of the United States National Museum 47. Washington, D.C.: GPO, 1896.

Jordan, David S., and Charles H. Gilbert. *Synopsis of the Fishes of North America.* Bulletin of the United States National Museum 16. Washington, D.C.: GPO, 1882.

Jørgensen, Dolly. *Recovering Lost Species in the Modern Age: Histories of Longing and Belonging.* Cambridge, MA: MIT Press, 2019.

Karam, Abraham Patrick. "History and Development of Refuge Management for Devils Hole Pupfish (*Cyprinodon diabolis*) and an Ecological Comparison of Three Artificial Refuges." Master's thesis, Southern Oregon University, 2005.

Keiter, Robert B. *To Conserve Unimpaired: The Evolution of the National Park Idea.* Washington, D.C.: Island Press, 2013.

Kelly, Isabel T., and Catherine S. Fowler. "Southern Paiute." In *Handbook of North American Indians.* Vol. 11, *Great Basin,* edited by Warren L. D'Azevedo, 368–397. Washington, D.C.: Smithsonian, 1986.

Kennedy, Stephen E. "Life History of the Leon Springs Pupfish, *Cyprinodon bovinus*." *Copeia* 1977, no. 1 (March 16, 1977):93–103.

Kingsland, Sharon E. *The Evolution of American Ecology, 1890–2000*. Baltimore: Johns Hopkins University Press, 2005.

Knack, Martha C. *Boundaries Between: The Southern Paiutes, 1775–1995*. Lincoln: University of Nebraska Press, 2001.

Koerner, Lisbet. *Linnaeus: Nature and Nation*. Cambridge, MA: Harvard University Press, 1999.

Kohler, Robert E. *All Creatures: Naturalists, Collectors, and Biodiversity, 1850–1950*. Princeton, NJ: Princeton University Press, 2006.

——. *Landscapes and Labscapes: Exploring the Lab-Field Border in Biology*. Chicago: University of Chicago Press, 2002.

Kolbert, Elizabeth. *The Sixth Extinction: An Unnatural History*. New York: Henry Holt, 2014.

——. *Under a White Sky: The Nature of the Future*. New York: Crown, 2021.

Lagler, Karl F., John E. Bardach, and Robert R. Miller. *Ichthyology*. New York: Wiley, 1962.

La Rivers, Ira. "The Dryopoidea Known or Expected to Occur in the Nevada Area (Coleoptera)." *Wasmann Journal of Biology* 8, no. 1 (1950):97–111.

——. *Fishes and Fisheries of Nevada*. Nevada State Fish and Game Commission, 1962. Reprint, Reno: University of Nevada Press, 1994.

La Rivers, Ira, and T. J. Trelease. "An Annotated Checklist of the Fishes of Nevada." *California Fish and Game* 38, no. 1 (1952):113–123.

Leakey, Richard, and Roger Lewin. *The Sixth Extinction: Patterns of Life and the Future of Humankind*. New York: Doubleday, 1995.

Lee, Ronald F. *The Antiquities Act of 1906*. Washington, D.C.: National Park Service, 1970.

Lema, Sean C. "Hormones, Developmental Plasticity, and Adaptive Evolution: Endocrine Flexibility as a Catalyst for 'Plasticity-First' Phenotypic Divergence." *Molecular and Cellular Endocrinology* 502 (2020):110678.

Lema, Sean C., and Gabrielle A. Nevitt. "Testing an Ecophysiological Mechanism of Morphological Plasticity in Pupfish and Its Relevance to Conservation Efforts for Endangered Devils Hole Pupfish." *Journal of Experimental Biology* 209 (2006):3499–3509.

Linduska, J. P. "Endangered Species—The Federal Program." *The Elepaio: Journal of the Hawaii Audubon Society* 29, no. 1 (July 1968):1–5.

Lingenfelter, Richard E. *Death Valley and the Amargosa: A Land of Illusion*. Berkeley: University of California Press, 1986.

Linnaeus, Carolus [Karl von Linné]. *Systema Naturae*, edited by M. S. J. Engel-Ledoboer and H. Engel. 1735. Reprint, Nieuwkoop, NL: B. de Graff, 1964.

Loeltz, O. J., Burdge Irelan, J. H. Robison, and F. H. Olmsted. *Geohydrologic Reconnaissance of the Imperial Valley, California*. U.S. Geological Survey Professional Paper 486-K. Washington, D.C.: GPO, 1975.

Lugaski, Thomas P. "Obituary: Ira John La Rivers, II, 1915–1977." *Pan-Pacific Entomologist* 55, no. 3 (1979):230–233.

Lyons, Lindsey Treon. "Temporal and Spatial Variation in Larval Devils Hole Pupfish (*Cyprinodon diabolis*) Abundance and Associated Microhabitat Variables in Devils Hole, Nevada." Master's thesis, Southern Oregon University, 2005.

Mackintosh, Barry. *The National Parks: Shaping the System*. Washington, D.C.: U.S. Department of the Interior, 1985.

Maienschein, Jane. "Pattern and Process in Early Studies of Arizona's San Francisco Peaks." *BioScience* 44, no. 7 (1994):479–485.

Marcot, Bruce G., and Jack Ward Thomas. *Of Spotted Owls, Old Growth, and New Politics: A History since the Interagency Scientific Committee Report.* General Technical Report PNW-GTR-408. U.S. Forest Service, Pacific Northwest Research Station, September 1997.

Martin, Andrew P., and Jennifer L. Wilcox. "Evolutionary History of Ash Meadows Pupfish (Genus *Cyprinodon*) Populations Inferred Using Microsatellite Markers." *Conservation Genetics* 5 (2004):769–782.

Martin, Andrew, P., Anthony A. Echelle, Gerard Zegers, Sherri Baker, and Connie L. Keeler-Foster. "Dramatic Shifts in the Gene Pool of a Managed Population of an Endangered Species May Be Exacerbated by High Genetic Load." *Conservation Genetics* 13 (2012):349–358.

Martin, Christopher H., Jacob E. Crawford, Bruce J. Turner, and Lee H. Simons. "Diabolical Survival in Death Valley: Recent Pupfish Colonization, Gene Flow, and Genetic Assimilation in the Smallest Species Range on Earth." *Proceedings of the Royal Society B: Biological Sciences* 283, no. 1823 (2016):20152334. doi.org/10.1098/rspb.2015.2334.

Martin, Christopher H., and Sebastian Höhna. "New Evidence for the Recent Divergence of Devils Hole Pupfish and the Plausibility of Elevated Mutation Rates in Endangered Taxa." *Molecular Ecology* 27 (2018):831-838. doi.org/10.1111/mec.14404.

Martin, Christopher H., Sebastian Höhna, Jacob E. Crawford, Bruce J. Turner, Emilie J. Richards, and Lee H. Simons. "The Complex Effects of Demographic History on the Estimation of Substitution Rate: Concatenated Gene Analysis Results in No More Than Twofold Overestimation." *Proceedings of the Royal Society B: Biological Sciences* 284, no. 1860 (2017):2017 0537. doi.org/10.1098/rspb.2017.0537.

Marx, Karl. "The Eighteenth Brumaire of Louis Napoleon." 1852. In *Karl Marx: Selected Writings*, edited by David McLellan, 329-355. New York: Oxford University Press, 2000.

Massie, Michael A. "The Defeat of Assimilation and the Rise of Colonialism on the Fort Belknap Reservation, 1875–1925." *American Indian Culture and Research* 7, no. 4 (1984):33–49.

McCallum, Malcolm L. "Vertebrate Biodiversity Losses Point to a Sixth Mass Extinction." *Biodiversity and Conservation* 24, no. 10 (2015):2497–2519. doi.org/10.1007/s10531-015-0940-6.

McGuire, Virginia L. *Water-Level and Recoverable Water in Storage Changes, High Plains Aquifer, Predevelopment to 2015 and 2013–15*. Scientific Investigations Report 2017-5040. Reston, VA: U.S. Geological Survey, 2017.

McWilliams, Carey. *Factories in the Field: The Story of Migratory Farm Labor in California*. 1939. Reprint, Berkeley: University of California Press, 1999.

Meine, Curt, Michael Soulé, and Reed F. Noss. "'A Mission-Driven Discipline': The Growth of Conservation Biology." *Conservation Biology* 20, no. 3 (2006):631–651.

Merriam, C. Hart. *Results of a Biological Survey of the San Francisco Mountain Region and Desert of the Little Colorado, Arizona*. North American Fauna 3. Washington, D.C.: GPO, 1890.

Miller, Robert Rush. "Coevolution of Deserts and Pupfishes (Genus Cyprinodon) in the American Southwest." In *Fishes in North American Deserts*, edited by Robert J. Naiman and David L. Soltz, 39–94. New York: Wiley, 1981.

———. "Correlation between Fish Distribution and Pleistocene Hydrography in Eastern California and Southwestern Nevada, with a Map of the Pleistocene Waters." *Journal of Geology* 54, no. 1 (1946):43–53.

———. *The Cyprinodont Fishes of the Death Valley System of Eastern California and Southwestern Nevada*. University of Michigan Museum of Zoology Miscellaneous Publications 68. Ann Arbor: University of Michigan Press, 1948.

———. "Records of Some Native Freshwater Fishes Transplanted into Various Waters of California, Baja California, and Nevada." *California Fish and Game* 54, no. 3 (1968):170–179.

———. "*Rhinichthys deaconi*, a New Species of Dace (Pisces: Cyprinidae) from Southern Nevada." Occasional Papers of the Museum of Zoology, University of Michigan 707. July 24, 1984.

———. "Speciation in Fishes of the Genera *Cyprinodon* and *Empetrichthys*, Inhabiting the Death Valley Region." *Evolution* 4, no. 2 (1950):155–163.

———. "Speciation Rates in Some Fresh-Water Fishes of Western North America." In *Vertebrate Speciation: A University of Texas Symposium*, edited by W. F. Blair, 537–560. Austin: University of Texas Press, 1961.

———. "The Status of *Cyprinodon macularius* and *Cyprinodon nevadensis*, Two Desert Fishes of Western North America." University of Michigan Museum of Zoology Occasional Papers 473. Ann Arbor: University of Michigan Press, 1943.

Miller, Robert Rush, Clark Hubbs, and Frances H. Miller. "Ichthyological Exploration of the American West: The Hubbs-Miller Era, 1915–1950." In *Battle Against Extinction: Native Fish Management in the American West*, edited by W. L. Minckley and James E. Deacon, 19–40. Tucson: University of Arizona Press, 1991.

Miller, Robert Rush, James D. Williams, and Jack E. Williams. "Extinction of North American Fishes during the Past Century." *Fisheries* 14, no. 6 (1989):22–38.

Minckley, W. L., and James E. Deacon, eds. *Battle against Extinction: Native Fish Management in the American West.* Tucson: University of Arizona Press, 1991.

Minteer, Ben A., James P. Collins, Karen E. Love, and Robert Puschendorf. "Avoiding (Re)extinction." *Science* 344 (April 18, 2014):260–261. doi.org/10.1126/science.1250953.

Muhn, James, and Hanson R. Stuart. *Opportunity and Challenge: The Story of* BLM. Washington, D.C.: GPO, 1988.

Murchison, Kenneth M. *The Snail Darter Case:* TVA *versus the Endangered Species Act.* Lawrence: University Press of Kansas, 2007.

Murphy, Bob. *Desert Shadows: A True Story of the Charles Manson Family in Death Valley.* Morongo Valley, CA: Sagebrush Press, 1993.

National Park Service. *Death Valley National Monument, California.* Washington, D.C.: GPO, 1934. https://www.nps.gov/parkhistory/online_books/brochures/1934/deva/1934.pdf.

Neuharth, Kayla D. "Modeling the Impacts of Sub-Seasonal Environmental Variability on the Devils Hole Pupfish." Master's thesis, University of Nevada, Reno, 2019.

Neumann, Roderick P. "Life Zones: The Rise and Decline of a Theory of the Geographic Distribution of Species." In *Spatializing the History of Ecology: Sites, Journeys, Mappings*, edited by Raf de Bont and Jens Lachmund, 37–55. New York: Routledge, 2017.

Nevada Test Site Community Act of 1963: Hearings on H.R. 8003 and S. 2030, Part 1, Before the Subcommittee on Legislation and Subcommittee on Communities of the Joint Committee on Atomic Energy, 88th Cong. 1963.

Norment, Christopher J. *Relicts of a Beautiful Sea: Survival, Extinction, and Conservation in a Desert World.* Chapel Hill: University of North Carolina Press, 2014.

Norris, Kenneth S. "To Carl Leavitt Hubbs, a Modern Pioneer Naturalist on the Occasion of his Eightieth Year." *Copeia* 1974, no. 3 (October 18, 1974):581–594.

Palmer, T. S. "National Monuments as Wild-Life Sanctuaries." In *Proceedings of the National Parks Conference*, 208–225. Washington, D.C.: GPO, 1917.

———, ed. *Place Names of the Death Valley Region in California and Nevada.* 1948. https://hdl.handle.net/2027/uc1.31822035077684.

Parker, Laura. "What Happens to the U.S. Midwest When the Water's Gone?" *National Geographic*, August 2016. https://www.nationalgeographic.com/magazine/2016/08/vanishing-midwest-ogallala-aquifer-drought/.

Pellegrin, Pierre. *Aristotle's Classification of Animals: Biology and the Conceptual Unity of the Aristotelian Corpus.* Translated by Anthony Preus. Berkeley: University of California Press, 1986.

Pisani, Donald J. "Federal Reclamation and Water Rights in Nevada." *Agricultural History* 51, no. 3 (1977):540–558.

——. *Water, Land, and Law in the West: The Limits of Public Policy, 1850–1920.* Lawrence: University Press of Kansas, 1996.

Pister, Edwin P. "The Desert Fishes Council: Catalyst for Change." In *Battle Against Extinction: Native Fish Management in the American West*, edited by W. L. Minckley and James E. Deacon, 55–68. Tucson: University of Arizona Press, 1991.

——, ed. *Proceedings of the Desert Fishes Council.* Vols. 3–4. Bishop, CA: Desert Fishes Council, 1971–1972 (December 1983). http://www.desertfishes.org.

——. *The Rare and Endangered Fishes of the Death Valley System—A Summary of the Proceedings of a Symposium Relating to their Protection and Preservation.* Bishop, CA: Desert Fishes Council, 1970.

——. "Species in a Bucket." *Natural History* 102, no. 1 (1993):14–18.

——. "Wilderness Fish Stocking: History and Perspective." *Ecosystems* 4, no. 4 (2001):279–286.

Pittman, Craig. *Cat Tale: The Wild, Weird Battle to Save the Florida Panther.* Toronto: Hanover Square Press, 2020.

Powell, John Wesley. *Report on the Lands of the Arid Region of the United States.* 2nd edition. Washington, D.C.: GPO, 1879.

Price, Jennifer. *Flight Maps: Adventures with Nature in Modern America.* New York: Basic Books, 1999.

Pritchard, James A. "The Meaning of Nature: Wilderness, Wildlife, and Ecological Values in the National Parks." *George Wright Forum* 19, no. 2 (2002):46–56.

Propst, David L., Jack E. Williams, Kevin R. Bestgen, and Christopher W. Hoagstrom, eds. *Standing between Life and Extinction: Ethics and Ecology of Conserving Aquatic Species in North American Deserts.* Chicago: University of Chicago Press, 2020.

Pyne, Stephen J. *California: A Fire Survey.* Tuscon: University of Arizona Press, 2016.

Reed, J. Michael, and Craig A Stockwell. "Evaluating an Icon of Population Persistence: The Devils Hole Pupfish." *Proceedings of the Royal Society B: Biological Sciences* 281, no. 1794 (2014). doi.org/10.1098/rspb.2014.1648.

Reed, Nathaniel P., and Dennis Drabelle. *The United States Fish and Wildlife Service.* Boulder, CO: Westview Press, 1984.

Reisner, Marc. *Cadillac Desert: The American West and Its Disappearing Water.* Revised edition. New York: Penguin, 1993.

Riggs, Alan C., and James E. Deacon. "Connectivity in Desert Aquatic Ecosystems: The Devils Hole Story." In *Conference Proceedings, Spring-Fed Wetlands: Important Scientific and Cultural Resources of the Intermountain Region, 7–9 May 2002*, edited by D. W. Sada and S. E. Sharpe. DRI Report 41201. Reno: Desert Research Institute, 2004.

Ritvo, Harriet. *The Platypus and the Mermaid and Other Figments of the Classifying Imagination.* Cambridge, MA: Harvard University Press, 1997.

Roberts, Heidi, and Richard V.N. Ahlstrom, eds. *A Prehistoric Context for Southern Nevada.* Archaeological Report 11-05. Las Vegas: HRA Inc., 2012.

Rodríguez-Estrella, Ricardo, and Ma Carmen Blázquez Moreno. "Rare, Fragile Species, Small Populations, and the Dilemma of Collections." *Biodiversity and Conservation* 15 (2006):1621–1625. doi.org/10.1007/s10531-004-4308-6.

Roman, Joe. *Listed: Dispatches from America's Endangered Species Act.* Cambridge, MA: Harvard University Press, 2011.

Rome, Adam. *The Genius of Earth Day: How a 1970 Teach-In Unexpectedly Made the First Green Generation.* New York: Hill and Wang, 2013.

Rosner, Hillary. "Attack of the Mutant Pupfish." *Wired,* November 19, 2012. https://www.wired.com/2012/11/mf-mutant-pupfish/.

Rothman, Hal K. *Neon Metropolis: How Las Vegas Started the Twenty-First Century.* New York: Routledge, 2002.

———. *Preserving Different Pasts: The American National Monuments.* Urbana: University of Illinois Press, 1989.

Rothman, Hal K., and Char Miller. *Death Valley National Park: A History.* Reno: University of Nevada Press, 2013.

Runck, Clay, and Dean W. Blinn. "Role of *Belostoma bakeri* (Heteroptera) in the Trophic Ecology of a Fishless Desert Spring." *Limnology and Oceanography* 39, no. 8 (1994):1800–1812.

Runte, Alfred. *National Parks: The American Experience.* 3rd edition. Lincoln: University of Nebraska Press, 1997.

Sabin, Paul. "Environmental Law and the End of the New Deal Order." *Law and History Review* 33, no. 4 (2015):965–1003.

Sackett, Joshua D., Desiree C. Huerta, Brittany R. Kruger, Scott D. Hamilton-Brehm, and Duane P. Moser. "A Comparative Study of Prokaryotic Diversity and Physiochemical Characteristics of Devils Hole and the Ash Meadows Fish Conservation Facility, a Constructed Analog." *PloS ONE* 13, no. 3 (2018):e0194404.

Sada, Donald W. *Recovery Plan for the Endangered and Threatened Species of Ash Meadows, Nevada.* Portland, OR: U.S. Fish and Wildlife Service, 1990.

Sağlam, İsmail K., Jason Baumsteiger, and Michael R. Miller. "Failure to Differentiate between Divergence of Species and Their Genes Can Result in Over-estimation of Mutation Rates in Recently Diverged Species." *Proceedings of the Royal Society B: Biological Sciences* 284, no. 1860 (2017):20170021. doi.org/10.1098/rspb.2017.0021.

Sağlam, İsmail K., Jason Baumsteiger, Matt J. Smith, Javier Lineres-Casenave, Andrew L. Nichols, Sean M. O'Rourke, and Michael R. Miller. "Best Available Science Still Supports an Ancient Common Origin of Devils Hole and Devils Hole Pupfish." *Molecular Ecology* 27 (2018):839–842. doi.org/10.1111.mec.14502.

———. "Phylogenetics Support an Ancient Common Origin of Two Scientific Icons: Devils Hole and Devils Hole Pupfish," *Molecular Ecology* 25 (2016):3962. doi.org/10.1111/mec.13732.

Sasser, Jim. "Learning from the Past: The Senate Code of Conduct in Historical Perspective." *Cumberland Law Review* 8 (1977):357–384.

Schamberger, Hugh A. *Evolution of Nevada's Water Laws, as Related to the Development and Evolution of the State's Water Resources, from 1866 to about 1960.* Water-Resources Bulletin 46. U.S. Geological Survey and Nevada Division of Water Resources, 1991.

Seawright, Jason, and John Gerring. "Case Selection Techniques in Case Study Research:

A Menu of Qualitative and Quantitative Options." *Political Research Quarterly* 61, no. 2 (June 2008):294–308.

Sellars, Richard West. *Preserving Nature in the National Parks: A History*. New Haven, CT: Yale University Press, 1997.

Shapovalov, Leo, and William A. Dill. "A Check List of the Fresh-Water and Anadromous Fishes of California." *California Fish and Game* 36, no. 4 (1950):382–391.

Shnayerson, Michael. "Who's Ruining Our National Parks?" *Vanity Fair*, October 17, 2006. http://www.vanityfair.com/news/2006/06/nationalparks200606.

Shor, Elizabeth N., Richard H. Rosenblatt, and John D. Isaacs. *Carl Leavitt Hubbs, 1894–1979: A Biographical Memoir*. Washington: National Academy of Sciences, 1987.

Shurts, John. *Indian Reserved Water Rights: The Winters Doctrine in Its Social and Legal Context, 1880s–1930s*. Norman: University of Oklahoma Press, 2000.

Skillen, James R. *The Nation's Largest Landlord: The Bureau of Land Management in the American West*. Lawrence: University Press of Kansas, 2009.

Skowronek, Stephen. *Building a New American State: The Expansion of National Administrative Capacities, 1877–1920*. Cambridge: Cambridge University Press, 1982.

Soulé, Michael E., and Bruce A. Wilcox, eds. *Conservation Biology: An Evolutionary-Ecological Perspective*. Sunderland, MA: Sinauer Associates, 1980.

Smith, Michael L. *Pacific Visions: California Scientists and the Environment, 1850–1915*. New Haven, CT: Yale University Press, 1987.

Spence, Mark David. *Dispossessing the Wilderness: Indian Removal and the Making of the National Parks*. New York: Oxford University Press, 1999.

Stegner, Wallace. *Beyond the Hundredth Meridian: John Wesley Powell and the Second Opening of the West*. Boston: Houghton Mifflin, 1954. Reprint, New York: Penguin, 1992.

Stein, Bruce A., Lynn S. Kutner, and Jonathan S. Adams. *Precious Heritage: The Status of Biodiversity in the United States*. New York: Oxford University Press, 2000.

Stein, Judith. *Pivotal Decade: How the United States Traded Factories for Finance in the Seventies*. New Haven, CT: Yale University Press, 2010.

Sterling, Keir B. "Builders of the U.S. Biological Survey, 1885–1930." *Journal of Forest History* 33, no. 4 (1989):180–187.

———. *Last of the Naturalists: The Career of C. Hart Merriam*. New York: Arno Press, 1974.

"Summary of the Meetings 1976." *Copeia* 1976, no. 4 (1976):848–858.

Sumner, Francis B. "The Need for a More Serious Effort to Rescue a Few Fragments of Vanishing Nature." *The Scientific Monthly* 10, no. 3 (March 1920):236–248.

———. "The Responsibility of the Biologist in the Matter of the Preserving Natural Conditions." *Science* 54, no. 1385 (July 15, 1921):39–43.

Sumner, F. B., and M. C. Sargent. "Some Observations on the Physiology of Warm Spring Fishes." *Ecology* 21, no. 1 (January 1940):45–54.

Sumner, Lowell. "Biological Research and Management in the National Park Service: A History." *George Wright Forum* 3, no. 4 (1983):3–27.

Svancara, Leona K., J. Michael Scott, Dale D. Goble, Frank W. Davis, and Donna Brewer. "Endangered Species Time Line." In *The Endangered Species Act at Thirty*. Vol. 2, *Biodiversity in Human-Dominated Landscapes*, edited by J. Michael Scott, Dale D. Goble, and Frank W. Davis, 24–35. Washington, D.C.: Island Press, 2006.

Switzer, Jacqueline Vaughn. *Green Backlash: The History and Politics of Environmental Opposition in the U.S.* Boulder, CO: Lynne Rienner Publishers, 1997.

Taylor, Joseph E. *Making Salmon: An Environmental History of the Northwest Fisheries Crisis.* Seattle: University of Washington Press, 1999.

Taylor, Kathie. *Ash Meadows: Where the Desert Springs to Life.* Death Valley, CA: Death Valley Natural History Association, 2015.

Thomas, Craig W. *Bureaucratic Landscapes: Interagency Cooperation and the Preservation of Biodiversity.* Cambridge, MA: MIT Press, 2003.

Thomas, David Hurst, Lorann S.A. Pendleton, and Stephen C. Cappannari. "Western Shoshone." In *Handbook of North American Indians. Vol. 11, Great Basin,* edited by Warren L. D'Azevedo, 262–283. Washington, D.C.: Smithsonian, 1986.

Thomas, Jessica E., Gary R. Carvalho, James Haile, Nicolas J. Rawlence, Michael D. Martin, Simon Y.W. Ho, Arnór Þ Sigfússon, et al. "Demographic Reconstruction from Ancient DNA Supports Rapid Extinction of the Great Auk," *eLife* 8 (2019):e47509. doi.org/10.7554/eLife.47509.

Tjossem, Sara Fairbank. "Preservation of Nature and Academic Respectability: Tensions in the Ecological Society of America, 1915–1979." PhD diss., Cornell University, 1994.

Turner, James Morton. *The Promise of Wilderness: American Environmental Politics since 1964.* Seattle: University of Washington Press, 2012.

Turner, Thomas F., Thomas E. Dowling, Trevor J. Krabbenhoft, Megan J. Osborne, and Tyler J. Pilger. "Conservation Genetics of Desert Fishes in the Genomics Age." In *Standing between Life and Extinction,* edited by David L. Propst, Jack E. Williams, Kevin R. Bestgen, and Christopher W. Hoagstrom, 207–223. Chicago: University of Chicago Press, 2020.

U.S. Bureau of Reclamation. *Amargosa Project: California-Nevada, Concluding Report.* Boulder City, NV: U.S. Bureau of Reclamation, February 1975.

———. *Interim Report on Inland Basins Projects, Nevada-California, Amargosa Project: Reconnaissance Investigation,* Region 3. Boulder City, NV: U.S. Bureau of Reclamation, July 1968.

U.S. Department of Agriculture. *The Death Valley Expedition: A Biological Survey of Parts of California, Nevada, Arizona, and Utah,* part 2. North American Fauna No. 7. Division of Ornithology and Mammalogy, U.S. Department of Agriculture. Washington: GPO, 1893.

U.S. Department of Commerce. *1980 Census of Population.*

U.S. Department of Energy. *Atomic Power in Space: A History.* DOE/NE/32117-H1. Washington, D.C.: U.S. Department of Energy, 1987.

———. *United States Nuclear Tests: July 1945 through September 1992.* DOE/NV-209-REV 15. 2000.

U.S. Department of the Interior. *Annual Report of the Secretary of the Interior, Fiscal Year Ending June 30, 1946.* Washington, D.C.: GPO 1947.

Van Dyke, Fred. *Conservation Biology: Foundations, Concepts, Applications.* 2nd edition. Springer: 2008.

Vetter, Jeremy. *Field Life: Science in the American West during the Railroad Era.* Pittsburgh, PA: University of Pittsburgh Press, 2016.

Vidal, Fernando, and Nélia Dias, eds., *Endangerment, Biodiversity, and Culture.* New York: Routledge, 2016.

Vincent, Carol Hardy, Laura A. Hanson, and Carla N. Argueta. *Federal Land Ownership: Overview and Data.* Congressional Research Service Report R42346. Washington, D.C.: Library of Congress, March 3, 2017.

Vinyard, Gary, and James E. Deacon. Foreword to *Fishes and Fisheries of Nevada,* by Ira La Rivers, 1–4. Reno: University of Nevada Press, 1994.

Wales, Joseph H. "Biometrical Studies of Some Races of Cyprinodont Fishes, from the Death Valley Region, with Description of *Cyprinodon diabolis*, n. sp." *Copeia* 1930, no. 3 (September 30, 1930):61–70.

———. "Life History of the Blue Rockfish, *Sebastodes mystinus*." Master's thesis, Stanford University, 1932.

Walker, George E., and Thomas E. Eakin. *Geology and Ground Water of Amargosa Desert, Nevada–California*. Reconnaissance Series Report 14. Carson City: Nevada Department of Conservation and Natural Resources, 1963.

Walker, Richard A. *The Conquest of Bread: 150 Years of Agribusiness in California*. New York: New Press, 2004.

Webb, Walter Prescott. "The American West, Perpetual Mirage." *Harper's Magazine*, May 1, 1957, 25–31.

Weldon, Fred W. "History of Water Law in Nevada and the Western States." Background Paper 03-2. Legislative Counsel Bureau, Nevada Legislature, 2003. https://www.leg.state.nv.us/Division/Research/Publications/Bkground/BP03-02.pdf.

Wheatley, Charles F., Jr. *Study of Land Acquisitions and Exchanges Related to Retention and Management or Disposition of the Federal Public Lands*. Prepared for the Public Land Law Review Commission. Springfield, VA: Federal Scientific and Technical Information Service, 1970.

Wilbur, Sanford R. *The California Condor, 1966–76: A Look at Its Past and Future*. North American Fauna 72. Washington, D.C.: U.S. Department of the Interior, 1978.

Wilcove, David S., and Margaret McMillan. "The Class of '67." In *The Endangered Species Act at Thirty*. Vol. 1, *Renewing the Conservation Promise*, edited by Dale D. Goble, J. Michael Scott, and Frank W. Davis, 45–50. Washington: Island Press, 2006.

Wilkins, John S. *Species: A History of the Idea*. Berkeley: University of California Press, 2009.

Williams, J. E. *Observations on the Status of the Devils Hole Pupfish in the Hoover Dam Refugium*. Report REC-ERC-77-11. Denver, CO: U.S. Bureau of Reclamation, September 1977.

Williams, Jack E., and Donald W. Sada. "Ghosts of Our Making: Extinct Aquatic Species of the North American Desert Region." In *Standing between Life and Extinction, Ethics and Ecology of Conserving Aquatic Species in North American Deserts*, edited by David L. Propst, Jack E. Williams, Kevin R. Bestgen, and Christopher W. Hoagstrom, 89–105. Chicago: University of Chicago Press, 2020.

Williams, Owen R., Jeffrey S. Albright, Paul K. Christensen, William R. Hansen, Jeffrey C. Hughes, Alice E. Johns, Daniel J. McGlothlin, et al. "Water Rights and Devils Hole Pupfish at Death Valley National Monument." In *Science and Ecosystem Management in the National Parks*, edited by William L. Halvorson and Gary E. Davis, 161–183. Tucson: University of Arizona Press, 1996.

Wilson, Edward O. *The Diversity of Life*. Cambridge, MA: Harvard University Press, 1992.

Wilson, Kevin P., and Dean W. Blinn. "Food Web Structure, Energetics, and Importance of Allochthonous Carbon in a Desert Cavernous Limnocrene: Devils Hole, Nevada." *Western North American Naturalist* 67, no. 2 (2007):185–198.

Wilson, Kevin P., Mark B. Hausner, and Kevin C. Brown. "The Devils Hole Pupfish: Science in a Time of Crises." In *Standing between Life and Extinction, Ethics and Ecology of Conserving Aquatic Species in North American Deserts*, edited by David L. Propst, Jack E. Williams, Kevin R. Bestgen, and Christopher W. Hoagstrom, 379–389. Chicago: University of Chicago Press, 2020.

Wilson, Robert M. *Seeking Refuge: Birds and Landscapes of the Pacific Flyway.* Seattle: University of Washington Press, 2010.

Winograd, Isaac J., et al. "Continuous 500,000-Year Climate Record from Vein Calcite in Devils Hole, Nevada." *Science* 258, no. 5080 (October 9, 1992):255–260.

Winograd, Isaac J. *A Summary of the Ground-Water Hydrology of the Area between the Las Vegas Valley and the Amargosa Desert, Nevada, with Special Reference to the Effects of Possible New Withdrawals of Ground Water.* Report TEI-840. U.S. Geological Survey, September 1963.

Worster, Donald. *Dustbowl: The Southern Plains in the 1930s.* New York: Oxford University Press, 1979.

———. *Rivers of Empire: Water, Aridity, and the Growth of the American West.* New York: Oxford University Press, 1985.

Worts, G. F., Jr. "Effect of Ground-Water Development on the Pool Level in Devils Hole, Death Valley National Monument, Nye County, Nevada." Carson City, NV: U.S. Geological Survey, Water Resources Division, August 1963.

Wright, George M., Joseph S. Dixon, and Ben H. Thompson. *Fauna of the National Parks of the United States: A Preliminary Survey of Faunal Relations in National Parks.* Fauna Series 1. Washington, D.C.: GPO, 1933.

Wright, R. Gerald. *Wildlife Research and Management in the National Parks.* Urbana: University of Illinois Press, 1992.

Wulf, Andrea. *The Invention of Nature: Alexander von Humboldt's New World.* New York: Vintage, 2015.

Yochim, Michael J. *Protecting Yellowstone: Science and the Politics of National Park Management.* Albuquerque: University of New Mexico Press, 2013.

Yoon, Carol Kaesuk. *Naming Nature: The Clash between Instinct and Science.* New York: Norton, 2009.

Acknowledgments

I suspect that most research projects have humble beginnings. This book is no exception. In July 2013, I googled—in complete, blissful ignorance—the words "Devils Hole pupfish." That was my starting point. The distance I traveled from that first day to produce this book was a long one by many measures. I have, for example, now worked on it about seven times longer than the average Devils Hole pupfish lives, but it is perhaps best expressed in the many debts I have accumulated along the way.

I began writing about the pupfish after being hired by the American Society for Environmental History, which had just been contracted by the National Park Service to produce a report for biologists and other staff at Death Valley National Park on the pupfish's history. Lisa Mighetto, the society's former director, took a chance on me, and James Pritchard, the project's senior historian, joined in some early research and then helped me shape and edit drafts of the report. At Death Valley, cultural resources manager Blair Davenport and archivist Greg Cox oriented me to the park's library and archives. Aquatic ecologist Kevin P. Wilson introduced me to many aspects of Devils Hole science and provided numerous connections to other federal and state employees with knowledge (and archival papers) related to the pupfish. Park Service employees Mike Bower, Bailey Gaines, and Jeff Goldstein also offered invaluable assistance. The resulting report, *Recovering the Devils Hole Pupfish: An Environmental History*, was delivered to the Park Service in 2017. Although very different from this volume—in argument, length, prose, and intended audience—the report remains foundational for my knowledge of many issues discussed here.

As I wrapped up work on that report, I had the very good fortune to receive a fellowship from the University of California's Institute for the Study of Ecological and Evolutionary Climate Impacts. UC–Santa Barbara's Environmental Studies Program hosted me, and while there I was able to take time to reflect on the meaning of the pupfish story, conduct additional research, and draft several chapters. I doubt very seriously whether this book would have been written without the support I received at UC–Santa Barbara from Peter Alagona, Jennifer Martin, and Brian Tyrrell. All three now know more about pupfish than they ever bargained for and each dramatically improved the quality of this book. While working there, I also drafted three articles focused on different aspects of the pupfish's history. These articles, in *Forest History Today*, *Notes and Records*, and an edited collection, *Standing Between Life and Extinction* (coauthored with Kevin P. Wilson and Mark B. Hausner), are cited where appropriate.

Much of the evidence in this book comes from manuscript sources. I would not have been able to locate or access this material without the help of many archivists and collections managers at university and government repositories, especially Su Kim Chung, Christina Fidler, Michelle Koo, Kris Murphy, Deborah Osterberg, Heather Smedberg, Gerald Smith, and Jacquelyn Sundstrand.

I also received help in locating and accessing records held in government offices. At the U.S. Fish and Wildlife Service, Annji Bagozzi, Wendy Contrael, Corey Lee, Mark Pelz, Sharon McKelvey, Mark Meadows, Lee Simons, Richard Smith, and Darrick Weisenfluh all provided key support. Will Thomas, former hydrology technician at Ash Meadows National Wildlife Refuge, was the best tour guide I could have had during my first forays to the desert. At the Nevada Department of Wildlife, Kevin Guadalupe and John Sjöberg lent critical assistance. Employees at other agencies also located materials and offered advice, including Melanie Cota, Dan Randles, and Steve Parmenter. A number of others whose past work or activism connected them to Devils Hole took time to share stories and personal papers, including James E. Deacon, Gail Kobetich, Dave Livermore, Leontine (Tina) Nappe, Edwin P. (Phil) Pister, Don Sada, Doug Threloff, Roland Westergard, and Jack E. Williams.

I have also been very fortunate to learn through conversations and correspondence with historians and scientists who took time to discuss issues surrounding the pupfish, the history of science, or environmental history. They include especially Elena Aronova, Lincoln Bramwell, Scott Cooper, Anita Guerrini, Mark Hausner, Josh Howe, Andy Kirk, Dan Lewis, Patrick McCray, Char Miller, and John Soluri. At the University of Nevada Press, JoAnne Banducci and Margaret Dalrymple shepherded the manuscript to the finish line. Jeff Grathwohl provided expert copyediting.

Throughout this work, I have been sustained by my family and friends. My parents, June and Paul, caught the pupfish bug to the extent that, when they went on a cross-country road trip from New Jersey, they turned around and headed home after seeing Devils Hole and Death Valley. In addition to their general enthusiasm, I wrote and edited large chunks of the book at their home in the winter of 2019. How anyone writes a book who does not have a talented editor for a sibling is beyond me. My brother, Glenn M. Brown, provided expert criticism and advice at every stage. Margaret J. Krauss, Elin Ljung, and Paul McFarland each read drafts and commented on the manuscript. I finally met Nora Livingston in part thanks to the pupfish. She helped me finish this book, often by making me see how much there is to do in this world besides writing about a tiny fish.

In *Travels with Charley*, John Steinbeck observed that while planning a long journey, there is often a "private conviction" that the trip will not happen at all.* A similar kind of doubt followed me around while researching and writing this book. It was difficult to convince myself that the literal stacks of archival boxes of correspondence and reports, unread library books, notes, photographs, chapter drafts, and so on, would ever actually come together as a physical product. However, if you are reading this, then those doubts appear to have been unfounded; the book exists. What is useful and good about it is due in large measure to the people and institutions mentioned above. To all, thank you.

Lee Vining, California
January 2021

*John Steinbeck, *Travels with Charley in Search of America* (New York: Viking, 1962), 19.

Throughout this work I have been sustained by my family and friends. My parents [illegible] taught [illegible] to the extent that when they went [illegible] and [illegible] Death Valley. In addition [illegible] at [illegible] book [illegible] My brother [illegible] [illegible] and [illegible] Edith [illegible] [illegible] thanks to the [illegible] help. [illegible] book [illegible] there is [illegible] writing [illegible].

[illegible] the [illegible] that [illegible] will [illegible] [illegible] and [illegible] [illegible] books, notes, [illegible] [illegible] [illegible] the people and institutions mentioned above [illegible].

[illegible], California
[illegible]

Index

Page numbers in *italics* refer to figures.

About the Author

Kevin C. Brown works with his head and his hands on the east side of California's Sierra Nevada Mountains. He earned a doctorate in U.S. history at Carnegie Mellon University and was a postdoctoral researcher in the Environmental Studies Program at the University of California, Santa Barbara. He has also worked as a journalist and as a researcher for the National Park Service and the American Society for Environmental History. This is his first book.